Volume 11

THE OIL INDUSTRY AND GOVERNMENT STRATEGY IN THE NORTH SEA

THE OIL INDUSTRY AND GOVERNMENT STRATEGY IN THE NORTH SEA

ØYSTEIN NORENG

Routledge
Taylor & Francis Group

LONDON AND NEW YORK

First published in 1980 by Croom Helm Ltd

This edition first published in 2016
by Routledge
2 Park Square, Milton Park, Abingdon, Oxon OX14 4RN

and by Routledge
711 Third Avenue, New York, NY 10017

Routledge is an imprint of the Taylor & Francis Group, an informa business

British Library Cataloguing in Publication Data
A catalogue record for this book is available from the British Library

ISBN: 978-1-138-64127-3 (Set)
ISBN: 978-1-315-62232-3 (Set) (ebk)
ISBN: 978-1-138-65561-4 (Volume 11) (hbk)
ISBN: 978-1-138-65562-1 (Volume 11) (pbk)
ISBN: 978-1-315-62241-5 (Volume 11) (ebk)

Publisher's Note
The publisher has gone to great lengths to ensure the quality of this reprint but points out that some imperfections in the original copies may be apparent.

Disclaimer
The publisher has made every effort to trace copyright holders and would welcome correspondence from those they have been unable to trace.

THE OIL INDUSTRY AND GOVERNMENT STRATEGY IN THE NORTH SEA

ØYSTEIN NORENG

CROOM HELM LONDON

The International Research Center for Energy
and Economic Development (ICEED)
Boulder, Colorado

© 1980 Øystein Noreng
Reprinted 1983
Croom Helm Ltd, Provident House, Burrell Row,
Beckenham, Kent BR3 1AT

British Library Cataloguing in Publication Data

Noreng, Øystein
   The oil industry and government strategy in the North Sea
   1. Offshore oil industry – North Sea
   2. Energy policy – Great Britain
   3. Energy policy – Norway
   I. Title
   333.8'2   HD9575.N57
   ISBN 0-85664-850-7

Library of Congress Catalog Card Number: 80-81590

ISBN 0-918714-02-8

Copyright © 1980 Øystein Noreng

The International Research Center for Energy
and Economic Development (ICEED)
216 Economics Building
University of Colorado
Boulder, Colorado 80309 U.S.A.

Printed and bound in Great Britain by
Biddles Ltd, Guildford and King's Lynn

CONTENTS

TABLES

FIGURES

FOREWORD

For a number of years now, the International Research Center for Energy and Economic Development (ICEED) has carried out directly or sponsored research on economic and policy topics related to energy. Additionally, the Center has initiated an active publications programme. Aside from its *Journal of Energy and Development,* the ICEED publishes books, monographs, and proceedings of its annual international energy conference. Projects receiving the Center's attention range widely from case studies of the major energy-producing and exporting countries (with special emphasis on OPEC as well as such nations as Mexico) to the implications of capital surplus funds, the problem of absorptive capacity, the producer-consumer relationship, and the potential for international co-operation in energy.

The precarious nature of the global oil supply/demand balance was highlighted by the dislocations of 1979 arising from Iranian export cutbacks. Accordingly, the more 'secure' sources of Mexico, Venezuela, and Norway have come under closer scrutiny. Øystein Noreng's *The Oil Industry and Government Strategy in the North Sea* comes at a propitious time. Not only is the subject pressing, but the author is particularly suited to his task.

One of the most distinguished European energy analysts, Dr Noreng offers an admirable mix of academic and practical expertise, as evident in his earlier book *Oil Politics in the 1980s.* Moreover, his background and experience enable him to utilise a comprehensive approach, including both economic and political aspects, to delineate past policy rationale and the directions for future trends. The ICEED views the present volume as a major contribution to the fields of energy, resource management, and policy formulation.

Ragaei El Mallakh
Director, ICEED, and
Professor of Economics
University of Colorado
Boulder, Colorado

PREFACE

The initiative that I write a book on North Sea oil was originally taken
by Dr Ragaei El Mallakh, professor of economics at the University of
Colorado at Boulder and director of the International Research Center
on Energy and Economic Development (ICEED). During a visit to
Boulder in April 1977 there was agreement that I write a monograph on
British and Norwegian oil policies, provisionally to be entitled *The
Economics and Politics of North Sea Oil.* The title reflected both some
divergence of interests and a high level of ambition. The interest of
ICEED was mainly in an economic analysis, whereas my own interests
were more concerned with policy analysis. The ambition of both parties
was that the book be comprehensive on the subject. The time schedule
envisaged was likewise highly ambitious, the book was thought to
appear no later than early 1978.

As at that time I was busily working on another book project,
dealing with international oil matters, this work could only commence
early 1978, after the original deadline. It soon emerged that the book
would be larger than originally envisaged, and that it would focus more
on government policy than on the economics of the operations. Also, it
was clear that the book might also be of interest to a European public.
At this stage the publishers, Croom Helm in London, were brought into
the picture, and the project became a joint venture. Collection of data,
writing and rewriting of course took much longer time than anticipated,
and the final text was ready only by the summer of 1979. At this moment,
the dynamics of the world oil market were again on the move, changing
once more the premisses of the relationship between governments and
oil companies. Also, changes were coming up in British and Norwegian
oil policies, requiring more rewriting. In the meantime, the publication
of the book had been advertised on several occasions, causing some
frustration among prospective readers, for which I apologise.

The subject of the book is essentially the interdependence of
governments and oil companies in the extraction of a natural resource,
focusing on the choices and constraining factors of government policy
as well as on relevant administrative patterns. The perspective is largely
that of a government planner, whose main concerns are the long-term
and complex interests of the state, orderly development as well as social
and political stability. To some extent this reflects my own background,

as an economic planner with the Norwegian government and a planner with Statoil, Norway's national oil company. In addition, for some time I was on the board of the Norwegian Petroleum Workers' Union, NOPEF. This experience gave both useful insight and a valuable perspective.

I am indebted to a number of people for advice and helpful assistance. First of all, Ambassador Jens Evensen, Norway's former Minister of Law of the Sea, should be mentioned, both for fruitful working relationship in a study group on oil policy during 1979, and for his unique comparative study on petroleum legislation, *Oversikt over oljepolitiske spørsmål,* which has been a most useful reference work for the present study. Other Norwegian friends who have been helpful are Ingjald Ørbech Sørheim and Pål Erik Holte of the Ministry of Environmental Affairs, Johan Nic. Vold of the Ministry of Oil and Energy, Per Schreiner of the Planning Secretariat, Aage Frohde of the Ministry of Finance, Leif Ervik of the Christian Michelsen Institute, Martin Saeter of the Norwegian Institute of International Affairs, and many others. A special mention should be made of Dr Petter Nore of the Ministry of Oil and Energy, with whom I have enjoyed a stimulating working relationship over several years, and whose PhD thesis 'The Norwegian State's Relationship to the International Oil Companies over North Sea Oil, 1965–75' has been a most valuable work of reference. On the British side, Louis Turner of the Royal Institute of International Affairs, Paul Tempest of the Bank of England, Clive Jones of the Department of Energy, currently with the British embassy in Washington DC, should be mentioned, among several others. Thanks also go to the publishers, Ragaei El Mallakh of the ICEED and David Croom of Croom Helm, for encouragement and patience. Finally, I would like to thank the Norwegian School of Management and its rector Gerson Komissar for liberal working conditions and the Rockefeller Foundation's International Relations Division, especially Dr Mason Willrich and Dr Edwin A. Deagle, for generous support. Tom Wallin, now with Petroleum Intelligence Weekly, has edited part of the book and hopefully contributed to making language understandable. Lise Fogh and Jorunn Christensen of my faculty have been helpful, efficient and meticulous in typing the manuscript. Thomas Baumgartner of the Institute of Sociology, University of Oslo, has given valuable advice on the final version of the text.

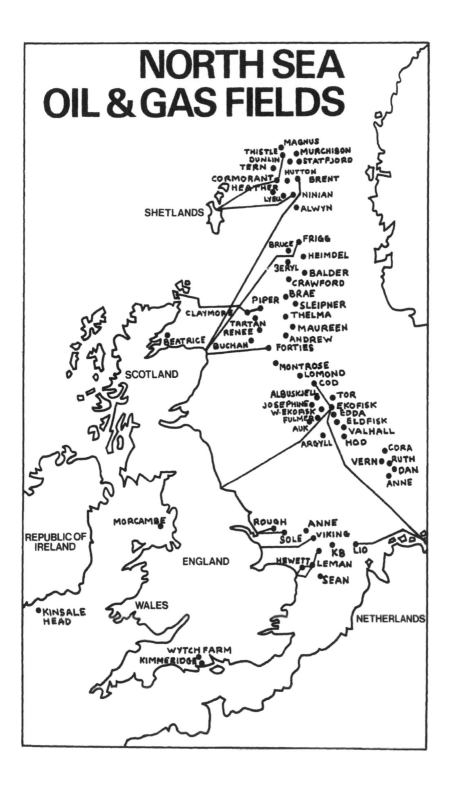

INTRODUCTION: THE RELATIONSHIP BETWEEN OIL COMPANIES AND GOVERNMENTS

This book deals with the relationship between oil companies and governments in oil producing countries, specifically in the UK and Norway. Because these two nations are among the world's youngest oil producing countries, the relationship between governments and oil companies here in many ways has been characterised by more uncertainty than in other oil producing countries. Even if the UK and Norwegian governments have long records of macro-economic and industrial policies, they had until recently quite limited knowledge of oil exploration, development and production. Thus they did not know the industry, and they could not fully appreciate the consequences of permitting the oil industry to develop locally. On the part of the oil companies there has been corresponding uncertainty. Although they had experience in oil exploration, production, refining and marketing, they had until recently only limited experience of operations in countries with governments actively intervening in economic life, so that they did not know the political environment into which they were moving and could not fully foresee the political responses that their presence would provoke. The history of the oil industry in the North Sea is to a large extent that of two separate learning processes, with the UK and Norwegian governments gradually learning the micro-economics of the oil industry, and becoming more able to exercise control, and with the oil industry gradually learning that operating in a UK or Norwegian context is quite different from that of working either in a North American context or in a developing country. Most probably, these learning processes are far from complete yet.

The relationship between oil companies and governments has undergone drastic changes in the latter part of the twentieth century. Until recently the pattern was set by the United States, Canada and a number of developing countries. Through the traditional concessionary system, oil companies had control of large areas of prospective land, with little government interference. In matters of exploration, development and depletion policy, oil companies were practically sovereign. In addition, the governments of the producing countries usually received a fairly small share of the rent of the oil, defined as the sum of incomes and profits. Furthermore, the control and taxation of the oil industry were

13

complicated by the fact that most of the oil was traded in integrated networks, with arm's-length transactions representing only a small share of the volume traded. It took the 'oil revolution' of the 1970s to change this pattern. The increase of the f.o.b. price of oil and the nationalisation of the oil industry in most OPEC countries have created a new situation, where the governments of the oil exporting countries are sovereign in matters of oil exploration, development and depletion policy, which is now being extended to oil trading and marketing, and where their share of the total oil rent is significantly larger than previously.

With hindsight it is easy to realise that the old pattern of relationship between oil companies and governments was hardly satisfactory from the point of view of the governments of the oil exporting countries. Indeed, it may be legitimate to wonder why and how it could persist as long as it did. The historical and therefore transitory nature of the old pattern of relationship between oil companies and governments was to some extent made clear by UK and Norwegian oil policies well before the 'oil revolution'. In the early 1960s, when first approached by the oil industry, both the UK and Norwegian governments decided that they could not accept the internationally predominant pattern of relationship, as embodied in the traditional form of the concessionary system. They would need more influence with exploration, development and depletion policies. At this time the North Sea oilfields were economically marginal at best, which gave the two governments a fairly weak bargaining position, particularly in taxation matters. Nevertheless the two governments elaborated their own versions of the concessionary system, with terms that from the outset were considerably stricter than those prevailing in almost all other oil producing countries, whether in North America or in developing regions of the world. This modified version of the concessionary system has often been referred to as the 'North Sea model'. Thus, two North-West European countries with essentially market economies and fairly moderate governments could in the 1960s appear as champions of oil policy radicalism, and in hindsight even as 'pre-revolutionaries'. However, this did not reflect any serious radicalism or willingness to overthrow the capitalist mode of production in either country. The oil policy radicalism in the UK and Norway simply reflected the fact that both countries were fairly mature democracies, with articulate interest groups and political parties which at a national level could systematically aggregate different social demands and formulate relevant policies. Part of this is an established tradition where governments are held responsible for social and economic problems, and have to act accordingly. This context of politically mature capitalism was new to most of the oil industry. In the United States, and to a lesser extent in Canada, the tradition has been for the government to stay out of economic life and in particular not to interfere with specific branches of industry. In the oil producing developing countries democracy has been incomplete and unstable, or non-existent.

This is also a question of dynamics. In the 1960s the UK and Norway had oil policies in advance of those of the OPEC countries. In the late 1970s they have been overtaken in matters of government control and government take. To some extent this is due to explicit policy choices, and to some extent it is due to incomplete learning.

The Uncertainty of Governments

In most capitalist industrial countries the governments have little experience of running businesses. Even in countries with a sizeable public industrial sector, as was the case in the UK especially, to a much lesser extent in Norway, there is usually an 'arm's length' relationship between government and nationalised industries, so that the public administration gains little insight into the micro-economic realities of the different branches of industry. Consequently, government policy is essentially of a macro-economic character, relying upon general tools of policy, with selective interventions being the exception, and often carried out without enough information. In relation to the international oil industry, the position of the UK and Norway had essentially been that of consumers and importers relying on the international oil companies to supply the domestic market with foreign oil. Consequently, a considerable part of the micro-economic realities of the international oil industry were not readily available to the UK and Norwegian governments. Against this background, the shift of the UK and Norway from being consumers of oil to becoming potential producers represented a qualitative change in the tasks of government of the two countries. The argument here is that the oil industry is different and more difficult to control from outside than most other industries, particularly when compared with the traditional industries of the UK and Norway. The reasons are the capital intensity of the oil industry, the historically high rates of profit, the difficulty of entering the industry and consequently the tendency towards joint ventures and reduction of competition, vertical integration, the traditionally low

price elasticities of supply and demand for oil, as well as the large cash
flows handled by oil companies. These factors distinguish the oil
industry from most other industries, making it more resistant to outside
interference. This has traditionally been the strength of the international
oil industry, and it has correspondingly been the weakness of governments. Historically, no single government of any oil importing country
has been able to control the oil companies, not even the United States.
The record of countries with large nationalised oil companies importing
foreign oil for domestic consumption does not indicate that government
control is readily exercised in this case either. For example, it is more
than doubtful whether the UK government is able to exercise any
effective control with British Petroleum, where it historically owns half
the shares. Correspondingly, the French and Italian governments do not
appear to have been very successful in controlling the operations of
their state oil companies.

For these reasons it is fair to assume that the UK and Norwegian
governments, when first approached by the international oil industry in
the early 1960s, did not know how to handle the oil companies. It is
equally fair to assume that the two governments did not know what it
would imply for the two economies and the two societies to become
producers and eventually net exporters of petroleum.

Evidently there was some understanding in the two governments that
the oil industry was difficult to control and that some special measures
were required. This is demonstrated by the provisions of the local
versions of the concessionary system, which aimed at forcing the oil
companies to explore and eventually to produce, as well as being aimed
at providing the governments with more information than was usually
the case with other industries, or in other oil producing countries.
However, with hindsight it appears that the two governments had little
knowledge of their own bargaining position in relation to the oil
companies. This is demonstrated by the gradual realisation that the
initial concessionary terms were not particularly favourable to the
governments. At least in Norway this led to terms becoming tougher
for the oil companies already before the 'oil revolution' of 1973–4.
The approach of the two governments was initially to treat the oil
industry more or less like any other industry, with a few exceptions. As
the governments received more knowledge about the national resource
bases, and about the oil industry itself, and were put under pressure
politically, it was realised that this general approach to problems was
quite unsatisfactory and could lead to negative results from the point of
view of government. Gradually the understanding has emerged in the

two governments that a more specific approach to oil matters was required. This gradual understanding has been the result of several factors, of increasing knowledge in government about the micro-economic realities of the oil industry, and of increasing political pressure and public debate on oil matters. Thus, the evolution of government policy in relation to oil has been linked to a learning process, with trials and errors. It has also been a question of adapting to new realities. As bureaucratic organisations with a bias for conservatism and keeping things as they are, this is not always the strength of governments. It has also been a question of being caught in the game, by establishing provisions and rules on a scarce base of information, but which later become institutionalised and thus difficult to alter. Political realities in North-West Europe are also that legislation, once enacted, is difficult to change, and that once a set of rules has been laid down, it to some extent gets a life of its own, regardless of changing circumstances.

The Uncertainty of Oil Companies

As already pointed out, until the 1960s the direct experience of the international oil industry as related to oil exploration and production was essentially confined to North America and a number of developing countries. There was no experience with large-scale oil production in North-West Europe. Technically and economically the North Sea was a new oil province, with problems of development and production qualitatively different from other oil provinces. Water depths and weather conditions make the North Sea, especially its northern parts, quite different from, for example, the Gulf of Mexico, not to mention the Middle East. In hindsight, given the delays and cost escalation on some projects, it appears that the oil companies did not have a full understanding of the problems involved in developing oilfields in the North Sea, especially in the northern parts (nor did governments, for that matter).

Politically, North-West Europe was a qualitatively new context for the international oil industry. In North America, the experience had been of governments that imposed a certain number of regulations, on the basis of anti-trust legislation, or on imports in order to protect the domestic producers from cheap imported oil, but that did not intervene directly in the industry. In the developing countries the experience had been of dealing with fairly weak governments, in many cases not representative of democratic processes and often actively supported by foreign interests, essentially emanating from unstable social and political systems, with a low level of education and limited expertise. In neither case did the oil companies have to cope with governments that were actively concerned with resources policy and prone to intervene directly in industry, nor with governments that could see the oil industry as a potential threat to macro-economic balances and that were prepared to restrict its activities for this reason, nor with strongly organised social interests actively influencing government to restrict the activities of the oil industry, nor with a strong and unified trade union movement. Against this background it is fair to assume that the oil companies did not know the political context into which they were moving. It is equally fair to assume that the oil companies did not know what issues could be raised politically in relation to their activities, and what solutions the political systems could work out.

The experience of the international oil industry in UK and Norwegian waters contains a number of apparent political surprises. In both countries there has been a general stiffening of terms for the oil companies, with increases of taxation, introduction of state participation, stricter legislation on labour conditions, safety, protection of the environment, etc. This has generally worked in the direction of reducing the sovereignty of the oil companies on their concessions, and of reducing profits. However, this has to a large extent been offset by the price rises on oil in the 1970s.

1 GOVERNMENTS AND COMPANIES

Antagonistic Interdependence

For the Western world the accelerating depletion of the most accessible conventional energy resources, such as coal in Western Europe and petroleum in North America, can be said to have two important effects on the political economy of energy. First, increasing Western dependence on foreign sources of energy, principally imported oil, creates demand pressure on world energy markets, and prepares the ground for price increases. Second, the Western search for energy increasingly moves into newer and less accessible areas in the Arctic and on the continental shelves, a process that is facilitated by price increases in the world energy market. To a large extent these new frontier areas are owned by governments,[1] which necessarily politicises the exploitation of these new energy sources.

The search for new energy resources has transformed the relationship between governments and the energy industry in the Western world. On the one hand, the governments possess substantial energy resources, but in most cases they lack the relevant expertise and technology, and are reluctant to put public capital into high-risk ventures. On the other hand, the energy industry has the relevant expertise and technology, and in most cases is more willing than governments to risk capital in new ventures, but it lacks the ownership of the resource potential. Consequently, governments and the energy industry find themselves in a bargaining situation. Neither side can do without the other, but they have diverging interests, perceptions and demands.

In this bargaining situation, both the energy industry and the governments are confronted with new challenges. The energy industry must evaluate the economic attraction, the potential return on capital, the costs, prices, future supply needs and risks of any project. The governments, as landowners and as representatives of the societies affected by the resource extraction, must elaborate a policy, organise a legal framework, and evaluate their freedom of action both in relation to the energy industry and in relation to their political system.

Within this context there are important feedbacks. Government taxation and regulation affect the economic attractiveness of the new resources to the energy industry, which in turn affects the bargaining

19

position of governments in relation to the energy industry. Ultimately, the economic attractiveness of the new resources to the energy industry can be regulated politically, but the freedom of action of governments in creating incentives or obstacles to resource extraction is limited by economic factors over which they have little or no control.

There are two basic issues in this bargaining between governments and the private energy industry: the division of the financial gains and control over the activity. The division of the financial gains determines the distribution of income between operators and landowners, i.e. the return on private capital and the government share. The control of the activity determines the ability of either party to plan ahead. For the energy industry control means the ability to dispose of available energy resources in a long-term perspective and to integrate the exploitation of the particular resource with other activities. For governments control means the ability to implement a rational resource policy and to safe-guard important national economic interests. The purpose of control for the government is to influence the micro-economic behaviour of the firms involved, so that it conforms more to national economic and social considerations than would have been the case without control. Thus, control has repercussions on the future revenues of the activity, both in terms of their absolute size and their partition between the government and the companies.

There is clearly a complex interrelationship between the aim of getting some part of the financial gains and the aim of control. Control can be a method of supervision that secures a given part of the revenue for the government. On the other hand, control can imply that the maximum potential share of the financial gains is not realised. The more complex the context of the economic activity is for the government, the more important are other considerations in relation to the realisa-tion of the maximum share of the financial gains, and the more control is a separate aim, distinct from the aim of income realisation.

The basic strategy problem for governments is that, to the extent that the resource extraction is not organised directly by governments, the delegation of operational tasks to private interests implies a certain degree of denationalisation of the resources.[2] This means that the operation has to be made financially attractive to the private energy industry, leading to a certain renunciation of economic rent and control. It is rational to delegate operational tasks to private interests that are more able than governments to organise the activity, but this difference in ability is also an obstacle to effective control. Further-more, the use of private capital implies that the government, explicitly or implicitly, accepts a given rate of return on the activity, which eventually can restrict the government's freedom of action.

For the energy industry, the basic problem of strategy is that the extraction of the energy resources cannot be seen primarily in the perspective of the energy market. First, government considerations are likely to be different from those of private companies, and must be taken into account. Second, if the context of the economic activity is complex, side-effects must be kept under control, otherwise unexpected constraints on the activity can arise. Thus, for private companies, the more complex the context, the larger the potential for external risks in relation to the economic activity. This means that the aim of maximising return on investment must be modified to a certain degree by other considerations.

In sum, there is as a result an interdependent relationship between governments and companies. It is not in the interest of governments, which have insufficient expertise and are unable or unwilling to risk public capital, that the private energy industry fares badly. Correspondingly, it is not in the interest of private energy industry that the governments run into serious conflicts in their domestic political systems or with public opinion. This is the objective basis for an alliance between governments and companies. But the alliance contains elements of conflict, as governments and companies are pursuing different objectives. Companies essentially want to maximise the return on capital and create conditions of long-term stability for themselves. This implies maximising their share of the financial gains and the control of the activity. Governments normally pursue a greater number of different objectives, but stability in external conditions can be assumed to have a high priority. Consequently, even governments very sympathetic to the demands of the energy industry are forced to make trade-offs with other considerations. In order to have an effective basis for making these trade-offs, governments also seek revenue and control in their handling of energy resources.

In this antagonistic relationship of interdependence, the ability to understand both the whole situation and the position of the other party is of prime importance in bargaining. The more governments understand about the energy market and the operations, motivations and calculations of the energy industry, the greater their chances of imposing their points of view on the companies. Correspondingly, the more the energy industry understands about the political context in which it is operating and the significance of governmental policies, motivations and considerations, the greater its chances of bargaining successfully. Knowledge to a certain extent can offset the initial weakness of each party in relation to the other.

The Oil Industry Moves into the North Sea

The general themes described above are illustrated in the North Sea context. In the history of the international oil industry the decision to move into the North Sea represented a significant change. At the time the decision was made, the estimated costs of production in the North Sea were perhaps twenty times those in the most favourable areas in the Middle East. Until then, the exploration and production of oil had been moving steadily into areas that were more accessible from a technical point of view. This created a consistent trend of declining marginal costs. In addition, the expansion of reserves, through discoveries and enhanced techniques of recovery, considerably exceeded the volume of oil produced and consumed each year. Given this background, the decision to move into the North Sea seems paradoxical.

However, the decision seems to have anticipated the shift that occurred in the world oil market around 1970, as the rapidly increasing volume of oil extracted and consumed surpassed a declining rate of reserve expansion. The pressure building up in the most accessible areas of production signified that, for technical and political reasons, a ceiling on the expansion of output in these areas was approaching. At a distance it seems that the catalyst was the change of oil policy in Libya, following the revolution of 1969, but this single event could not have had a substantial impact if it was not facilitated by more general conditions in the world oil market.

Since 1970 the trend of declining marginal costs seems to have reversed despite considerable short-term and medium-term fluctuations, and it is likely that the world will have to get increasing quantities of oil from the frontier areas, in the Arctic and on the continental shelves. In this light, the decision already taken around 1960 to go ahead in the North Sea appears to have been founded on a correct appreciation of long-term trends in the oil market.

The decision to move into the North Sea also had a particular political significance. Historically, the international oil industry extracted its oil in reasonably congenial political environments. In the United States the major political constraint has been the application of the anti-trust legislation. However, the oil industry has had important political support, and over a long period it enjoyed protection from the world oil market. After the revolution the international oil industry was expelled from the Soviet Union, the other major producer. Most other oil producers were developing countries, with a low level of economic development, a low level of education, an inexperienced civil service, and in most cases hardly a tradition of democratic government. In these countries the oil industry became a dominant force. The bargaining between governments and the industry has only taken place on equal terms very recently. As a result, oil policies tended to suit the needs of the international oil industry.

The UK and Norway around 1960 offered very different political contexts and from the outset were much less accommodating to the oil industry. Both countries had a highly developed and differentiated economy, a high level of education, an experienced civil service, and long traditions of democratic government. Both countries had a record of active government involvement in the economy. The oil industry could not expect a dominant position here, and the bargaining was going to be much tougher. The government bargainers could be expected to have greater freedom of action and better cards to play. Looking back, it seems evident that the decision to move into the North Sea in spite of high costs and a more cumbersome political context was motivated by an anticipation of much higher prices and of possible problems of supply on a global scale.

The decision to move into the North Sea also confronted the UK and Norwegian governments with problems of a new character. Oil initiated a process of political change in addition to its economic impact. In both countries oil has become increasingly politicised as the industry had its impact on the economy, social affairs and the environment. Furthermore, the increased economic surplus caused by the rise in the international price of oil has made public opinion demand tougher terms for the oil companies. At the same time the oil companies have increasingly accommodated themselves to the new situation.

Taking the whole period from the early 1960s to the late 1970s, it seems that the oil industry has suffered a setback in the North Sea. Costs have escalated and lead times increased. Taxation has been increased, state participation has been imposed, and there are increasingly tougher regulations concerning safety, protection of the environment and labour conditions. It could perhaps be argued that the economic attraction of the North Sea to the international oil industry has decreased, and it might be legitimate to ask why it is still there. The answer is that the economic attraction is still considerable, compared to most other oil producing areas, and that in any case it should be judged over a longer period of time. North Sea oil attracts some of the highest prices in the market, and company profits per barrel produced are several times higher than company profits on oil bought from OPEC countries' state oil companies. With prospects of relatively tighter oil supplies and higher prices during the rest of this century, North Sea oil appears to be a remarkably good deal for the companies. In addition, it is a fairly secure deal. This security obviously has a price, and this raises the question of a political rent for North Sea oil payable to the governments. The escalation of costs and lead times can at least partly be blamed on the oil industry itself, as it has had the main responsibility for technical operations.[3] The escalation of politically determined costs, such as taxation, state participation, regulations etc., is still fairly moderate in its overall effects. The relatively mature political contexts of the UK and Norway should only be seen as a disadvantage to the oil industry in a narrow sense. In the longer run, they may prove to be an advantage by offering security and continuity.

From the point of view of the two governments, the international oil industry's interest in the North Sea presented substantial advantages for economic policy and energy policy compared to the rest of Western Europe. There were also some obvious problems. Neither government had any direct experience with oil production nor much insight into the operations of the international oil industry. The oil industry has traditionally been difficult to control from outside[4] due to its integrated and non-competitive character.[5] This problem of control has been no less acute for the UK and Norwegian governments than for other governments in new oil producing areas.

The mature political contexts of the UK and Norway have been both an asset and a liability. With a stable and relatively well informed political base, the UK and Norwegian governments could take a tougher position in relation to the oil industry than could most governments of developing countries in the early 1960s.[6] Critical factors were the stability of the social and political systems, the level of education and information of the population, and the potential for foreign manipulation and domination.[7] However, the presence of the international oil industry, producing oil, has created a number of negative side-effects in the eyes of fairly important parts of public opinion. This has forced the two governments to anticipate problems and to take action. Oil policy has had to focus on both development and control, which has created a certain dualism in UK and Norwegian oil policy, leading to contradictions in attitudes and solutions.

For countries which are heavily dependent upon foreign oil, the UK and especially the Norwegian policy may appear as excessively prudent. However, the potential negative effects were real enough and, especially for Norway in the 1960s, the example of many other oil exporting countries was seen as a warning, as they were largely dominated by foreign oil companies, with limited possibilities of defending their own interests.[8] In addition, there were good practical reasons for caution. A major problem has been to keep the flow of money under control, in order to regulate money supply and the distribution of income, to avoid overheating the domestic economy, and to protect the balance of payments. It is a good sign of the political maturity of the UK and Norwegian contexts that both governments had to give macro-economic considerations top priority.

The fact that oil has been identified as a problem by the UK and Norwegian governments is less due to any enlightened idealism or to the presence of radical bureaucrats than to the way politics operate in the two countries. Both are welfare states, in which the public demands government provide certain basic services as well as a certain basic social protection, especially from potential man-made disasters. Any serious problems caused by the oil industry affecting large parts of the population would in the UK and Norway lead to protest being quickly and efficiently channelled back to the civil service through Parliament and the Cabinet. The cautious approach has been in the self-interest of the bureaucracies, pursuing stability in order to avoid political problems. This observation, including the fact that governments have given macro-economic considerations top priority, makes an interesting contrast with the orthodox Marxist notion that the state in a capitalist society unequivocally defends the interests of the best-organised capitalist forces, and the most profitable parts of private capital. It is evident that the international oil industry is better organised and more profitable than most, if not all, important parts of traditional UK and Norwegian capitalism. However, it is equally evident that the UK and Norwegian governments have not been tools of the private oil industry, in the sense that they have uncritically and directly advanced the interests of private profit at the expense of wider social and economic interests. In modern capitalist democracies, such as the UK and Norway, economic policy is at the forefront of the political process, and given the complexity of the economy as well as of domestic politics, the state first of all has to defend its own macro-economic interests, for the benefit of overall stability. The point is to manage the overall framework for the different kinds of economic activity and not to offend unduly important participants in the political process. This is why, when there has been a conflict between the profit interests of private oil business and the macro-economic interests of the state, the latter have generally won. This does not mean that the UK or the Norwegian state has fought the international oil industry, nor does it exclude a certain convergence of long-term interests between the oil industry and the state. However, given the complexity of the situation and the potential for political conflict over matters related to oil, co-operation between the oil industry and the civil service has focused on long-term issues; on short-term issues it is safer for the civil service to keep a certain distance in order not to compromise itself politically. Also, such co-operation on long-term issues can be eased and made politically more acceptable if the civil service keeps control in the short term. It is debatable whether this aspect of the game has been properly understood by all the private oil companies operating in UK and Norwegian waters. It suggests that on some critical issues the state in a capitalist society can understand the long-term interests of capitalism better than the capitalists themselves. It also means that the state defends the interests of the best-organised and most profitable parts of capitalism in a more indirect way than is often asserted by Marxist theory, and that the relationship between capitalist interests and public policy is far from linear. However, for the private oil companies operating in UK and Norwegian waters, the framework imposed by public policy is still fairly comfortable and the aggressiveness of the two governments in asserting their rights and duties of control has not been remarkable. This has also been the case with so-called leftist governments in the UK and Norway.[9]

Apart from their political determination to control the operations of the oil industry, the governments' problem is complicated by the fact that government agencies and oil companies are quite different types of organisation. They operate with different goals, with different resources and different premises; their relationship is based upon unequal terms. Classifying organisations by their value systems and goals, government agencies are institutions whose goals are related both to their future impact upon their environment and to their own future roles and resources, whereas oil companies are large business firms whose goals are related to their own future roles and resources. They both have different criteria for judging performance. The performance of a government agency cannot only be evaluated in quantitative terms; there must also be a qualitative element. The performance of an oil company, however, can be judged by quantitative yardsticks, such as growth and profits. This means that oil companies can focus on their own expansion, whereas government agencies have to take into account a number of different considerations. In addition, oil companies operate in a market, in an ever changing commercial and technical context, whereas the tasks of government agencies are usually highly stable. This implies different degrees of flexibility and different abilities in obtaining and treating information, giving oil companies a structural advantage over government agencies. This also means that the ability to cope with new problems is unequal.

The appearance of the international oil industry in UK and Norwegian waters presented the two governments with qualitatively new problems of control, because the operations differed from the traditional set of economic activities, and because the international oil industry is different from traditional UK and Norwegian industry. Organising a framework for controlling the oil industry and implementing the controls so that the oil industry could be successfully integrated into the existing economy has perhaps been the major task of domestic policy facing the UK and Norwegian governments in this century. It has required a high degree of sophistication in planning and in administration. Consequently, the UK and Norwegian oil policies should be examined at two levels: at the level of policy formulation and at the level of policy implementation. The first has to do with the identification of problems and the setting of priorities, the second has to do with the actual control of the operations. The UK and the Norwegian records have in many ways been remarkably different on these two points, reflecting a general difference in working traditions and procedures in the two administrations.

The Differences in Style between UK and Norwegian Oil Policy

Norwegian oil and related policies have generally been ambitious and sophisticated, emphasising theoretical argument and extensive documentation which have often received wide international attention. By contrast, UK oil and related policies have tended to be less ambitious and sophisticated, emphasising practical argument without much documentation.

Norwegian policies appear fairly ideological, as there is an explicit trade-off between values and interests, and explicit value preferences, while UK policies appear more pragmatic. In the implementation of policies, the Norwegian government is usually not very active and as a rule is not much engaged in matters of detail. According to Norwegian

administrative tradition, business is supposed to conform to public
policies and regulations on their own, without much supervision. The
UK administrative tradition is quite different: there is usually a more
detailed follow-up, with the civil service fairly actively ensuring that
policies and regulations are respected.[10] The Norwegian tradition of
administration appears remarkably non-bureaucratic, permitting a
combination of a fairly strong public direction of the economy with a
fairly independent economic life that is attentive and fairly loyal to
public policies.[11] The UK tradition appears as more bureaucratic,
leaving less to the free will of the economic agents, but also ensures a
more systematic control, even if the different policies may not be well
co-ordinated. The Norwegian system may be well suited for a small and
fairly homogeneous society, and the UK one may be better suited for a
large industrial country. The UK system of administration may also be
better suited for dealing with the international oil industry.

The notorious difficulty of controlling the international oil industry
from outside means that attempts by government to supervise its
operations do require a substantial administrative effort. At the level of
policy formulation, Norwegian oil policy has been characterised by
clarity, consistency and continuity, as well as by a certain degree of
boldness in relation to the international oil industry. However, at the
level of implementation, Norwegian policy has to abstain from detailed
supervision. UK oil policy at the level of formulation has tended to the
reverse, but in implementation the British have been more thorough in
their supervision. For the international oil companies, the Norwegians
may appear more hostile than the British in formulating policy, but in
practical implementation the British may well be tougher than the
Norwegians.

In formulating and implementing their oil policies the UK and
Norwegian governments have had to take a number of very different
considerations into account. In this relationship, the international oil
industry with all of its organisational and financial resources was but
one side of the problem. In both the UK and Norway matters related to
oil have become increasingly politicised as the oil industry has expanded
and the side-effects have become more visible. It would be, however,
erroneous to see the gradual toughening of the attitude of the UK and
Norwegian governments towards the international oil industry merely
as a function of increasing public pressure. The bureaucratic need for
stability and control manifested itself from the beginning of operations,
and recently the governments have taken steps to improve their
position. However, as long as the two governments have not mastered

the technology and are unwilling to risk capital, the North Sea operations will have to be kept attractive to private industry, which means that certain limits on government action must be respected.

For both governments oil policy represented a most formidable challenge to exploit and harness a new economic resource that few other countries possess. The UK and Norwegian governments have confronted decisions with multiple objectives, implying a clear definition of preferences and value trade-offs. The objectives have been relatively easy to formulate: improving the balance of payments, stimulating economic growth and industrial development, avoiding economic overheating, social and regional problems, etc. In fact, official documents on oil policy in both countries to a certain extent can be read as catalogues of good intentions and honourable aims. However, the major problem in practice is to define preferences explicitly and to make clear value trade-offs. This is a much more complicated task, and it is politically much more difficult, as certain social groups and economic interests will be less well treated than others, and in the UK and Norwegian political contexts this normally causes discomfort to government. The record so far is that the social and economic disaster that some observers feared has not occurred, but there have been many problems, and not all the potential benefits have been reaped. The oil policies of both countries can only be judged a moderate success. Given the enlarged economic potential offered by the oil, both countries had the possibility of embarking upon a strategy of balanced economic growth, improved balance of payments and modernising industrial plant and social and technical infrastructures, together with growth in private consumption. These different aims are certainly compatible in a dynamic perspective if reasonable proportions are maintained between the different uses of money. Given that oil in both cases is a finite resource, although it may be more finite for the UK than for Norway, two periods of time should be considered: during the period of oil extraction, both the UK and Norway have a considerable advantage compared to other Western European countries. In the period after, both countries could have an improved industrial base. In both cases, the risk is that the improvement of the industrial base will be sacrificed for greater private consumption. It appears that neither country is reaping all the potential long-term benefits, and that both may in fact have over-restricted their freedom of action in oil policy through economic policies. Part of the problem is the lack of adequate economic planning in both countries. In neither country do the economic and financial authorities appear to be really capable of handling long-term structural change. This is precisely the challenge, and this administrative failure could well become more acute in the future.

However, in all fairness, both the UK and Norwegian governments have managed to take account of a number of political considerations. Apart from macro-economic effects, both governments have been considering public revenue, industrial spin-off, regional, social and environmental effects. It seems that macro-economic considerations have been decisive in both countries, in particular with regard to depletion policy, which in many ways is the key element in oil policy and which strongly influences the bargaining position between governments and companies.[12] The difference in perceived absorptive capacity is relevant in this context. In the UK, unused productive capacity together with persistent problems of balance of payments have meant a great absorptive capacity. In Norway virtually full utilisation of the productive factors and a more satisfactory balance of payments have produced a small absorptive capacity. Consequently, the social rate of return for the two governments has differed quite considerably, being relatively high in the UK and relatively low in Norway. It could even be that the expected rate of increase in the price of oil falls between the social rates of return considered by the two governments. In this sense it has been rational for the UK to opt for a high rate of extraction and equally rational for Norway to opt for a low rate of extraction.

Through a high degree of continuity and stable political environments, UK and Norwegian oil policies can be said to accommodate the long-term interests of the oil industry, at some short-term cost. An essential element of this long-term accommodation is avoiding serious conflicts between the oil industry and its operational environment. Given the different sizes of the countries and the economies in relation to the potential oil resources, this risk of conflict is much greater in Norway than in the UK. For the UK oil is a very welcome stimulus to an industrial economy in deep financial and industrial trouble, with relatively little risk of dependence and dominance. For Norway, oil means an immense economic stimulus that must be harnessed in order to be beneficial, with a relatively high risk of dependence, dominance and uncontrolled social and economic change. This difference in scale, together with certain differences in the overall economic situation, in the government planning tradition and in the industrial base, explain a good deal of the differences in oil policies between the two countries.

Given these differences, it is hardly surprising that Norway sometimes seems to have given social, regional and environmental policy a higher priority than the UK. Not only is Norway better able to afford

such considerations, but there has been less conflict with the objectives
of the depletion policy.

However, macro-economic differences cannot be held responsible for
all differences in oil policy. There is also a question of method. The
elaboration of the oil policies in the UK and Norway can also be seen as
selective processes, where political systems and governments identify
tasks, give priorities and choose solutions. From this important
conclusions can be drawn as to how these two political systems function
in relation to new economic challenges, but also as to how the same
problems in the end often find similar solutions, even if circumstances
vary.

The North Sea Model

In developing their oil policies, the UK and Norwegian governments
have been in fairly close, but irregular, consultation. In global terms the
UK and Norwegian oil regimes are remarkably similar. On salient points in
the relationship between governments and companies, they seem to give
identical results. Together the UK and Norway have developed a new
model of resource management, which can be called the North Sea
model.

As the oil regimes were elaborated in the UK and Norway in the
early 1960s, they had to be based on existing models of resource
management,[13] at least as far as petroleum was concerned: the
concessionary model and the state model. The concessionary model was
based upon private companies being given *de facto* control over large
areas and being sovereign in questions of development and exploitation.
This model was mainly practised in North America and in most
developing countries, including most OPEC countries. The concessionary
model existed in two versions. In the United States concessions were
allocated by auction, in the rest of the world through administrative
decision.[14] In the state model the government of the country kept
control and organised the exploration and production itself, either
through an administrative agency, or a state oil company, or in
co-operation with private foreign partners, which received compensation
through service contracts or production sharing agreements. There were
also two versions of the state model: oil operations were directly
organised by government agencies in the Soviet Union, Romania and
China, or they were organised by a state oil company (a commercial
organisation operating in a market economy), as in Mexico or Indonesia.

In the early 1960s neither of these models was found fully satisfactory by the UK and Norwegian governments. Allocation of concessions
by auction was discarded, as it gave the governments too little power
over the distribution of licences. The concessionary system as practised
in most developing countries, especially in the OPEC countries, was
found to give companies an excessive dominance through generous
terms and large areas of concessions. The state system was discarded as
the two governments found they had insufficient knowledge and
experience to organise the oil operations themselves, and that in a
frontier area such as the North Sea, the experience of the international
oil industry was essential. Their solution was based on the concessionary
model, with administrative allocation of licences, but with stiffer terms
and much smaller areas of concessions than practised in most OPEC
countries. Nevertheless, the North Sea model at the outset had
structural features in common with most OPEC countries. Subsequently
the North Sea model has developed, in particular through the generalisation of state participation and through the introduction of additional
taxation.

In the late 1960s and early 1970s there has been a remarkable
structural development in the international oil industry, profoundly
affecting the relationship between governments and oil companies. As a
result, the concessionary model has been abandoned in most developing
countries, especially OPEC countries. Instead, the state model has been
introduced, not in its Soviet version, but in the version already practised
in Mexico and Indonesia. The main reasons for the OPEC countries
abandoning the concessionary model were that it was historically linked
with foreign dominance, it was found to give national governments
insufficient control, and it was considered detrimental to national
economic interests. Thus, the control of the national oil industry has
been seen as an important step in establishing national independence
and emancipation from a colonial or semi-colonial legacy.[15] The
dominant position of the national oil companies is a key feature of the
model of oil resource management which has spread in the 1970s. The
new oil regimes of countries such as Venezuela, Iraq, Algeria, Kuwait
and now also Saudi Arabia and Iran are typical examples. The national
oil company has, by law or in practice, a monopoly of exploration and
production. This usually gives a high degree of centralisation in
decision-making, with corresponding opportunities for long-term
planning of the operations, and for co-ordination of the various
activities.

The centralisation of the decision processes is an important feature in common with the Soviet version of the state model. The basic difference is that the 1970s version gives a key role to state oil companies (commercial organisations operating in a market economy, and generally in the world market). This version can be called the OPEC model, even if it is also practised outside OPEC. However, the key role played by the state oil companies provides similarity between the OPEC model and the North Sea model. Both can be described as state capitalist. However, the North Sea model usually permits private and foreign companies to participate directly in the exploration and the production, and it gives them a right to dispose of the oil, implying a certain denationalisation of the resource. Here is the major distinction from the OPEC model. Direct participation of private companies in exploration and production also implies a high degree of decentralisation in the decision processes, as participants vary between the different fields.

The differences and similarities can be seen from the matrix shown in Table 1.1.

Table 1.1: Models of Resource Management

Participation of Private Companies	Degree of Centralisation	Role of State Oil Companies	
		Dominant	Absent
Direct	Low	North Sea model	US model
Indirect	High	OPEC model	Soviet model

Both the North Sea model and the OPEC model were developed as alternatives to the traditional concessionary model practised in the United States. In the 1970s the political economy of international oil has changed fundamentally, with state participation or full nationalisation becoming the general pattern everywhere so far outside the United States. In this sense the North Sea model and the OPEC model represent a new political order in oil, with fairly similar aims, but with quite different organisation. The OPEC model, with its emphasis on exclusive state ownership, is antithetical to the previous order which gave foreign oil companies a dominant position. Most developing countries' enthusiasm for exclusive state ownership of the oil resources can be explained by their fear that national control is incompatible with direct foreign participation in exploration and production. The UK and

Norway never applied the traditional concessionary model and had less need to develop an opposite approach. In North-West Europe in the early 1960s it was politically feasible to elaborate a model of oil resource management that could accommodate private interests under public control. The advantage of direct participation by private foreign companies provided by the North Sea model is the access to the experience and technology of foreign partners, as well as the direct mobilisation of private capital. This perhaps makes the North Sea model more attractive and more relevant than the OPEC model for developed industrialised countries, whose economies are differentiated and whose political systems are stable. The attraction of the North Sea model to some other Western countries has been demonstrated, and Canada is a good example. The disadvantage of direct participation by private companies, and of the current state of the North Sea model, is that it yields the right to manage the operations and to dispose of the oil from the government to the private participants. This creates substantial problems of control. To offset this, the governments create a web of legal and financial regulations in order to capture a given part of the rent and in order to influence the micro-economic behaviour of the private companies. Thus, as political pressures build up and governments have to impose new regulations, the North Sea model becomes increasingly complex, with an increasing contradiction between the rights of the private concessionaries to manage the operations and to dispose of the oil, and the public controls. There is a risk of the North Sea model becoming increasingly economically inefficient, giving increasingly complex problems both of management and of control, and perhaps serving less and less well the interests of the governments, and maybe those of the companies as well. Although the North Sea model can be seen as a historical evolution of the concession-ary model, it is probably not the ultimate solution in oil resource management in North-Western Europe.

Historically, the participation of the international oil industry in the North Sea has been seen as a necessity because governments have been unable or unwilling to risk the capital and unable to organise operations. This implied a dependence upon private interests, reducing government freedom of action. Private profits had to be respected. At the micro-economic level control of operations is delegated to private interests. However, government keeps control of access to the resource and of political responsibility. This gives interdependence to public and private interests. This interdependence is largely antagonistic, because of conflict relating to the division of rent and to control of operations.

The basis of the interdependence is ever-changing, because of a dynamic international context and the experience of governments and companies. Thus, the interdependence is essentially unstable, containing the seeds of its own destruction, at least within the framework of the concessionary system.[16] This has particularly been the case in developing countries, where, historically, governments have been weak in relation to foreign oil companies.

In the North Sea context governments were initially in a stronger position. In Norway and the United Kingdom the concessionary system has been modified through active state participation, with national oil companies as the key element. Thus, the concessionary system as applied in Norway and the United Kingdom has never produced the same unequal relationship between government and foreign oil companies as has occurred in many developing countries. The national oil companies are also key elements in a national learning process, improving the ability to organise operations and, in the longer run, reducing dependence upon foreign expertise and foreign companies. Furthermore, experience indicates that the ability of the international oil industry to organise operations in the more difficult parts of the North Sea is not unlimited.[17] This could mean that the main argument of superior expertise for maintaining the presence of the international oil industry within the concessionary system is not entirely valid.

Notes

1. Mason Willrich, *Energy and World Politics* (Free Press, New York, 1975), pp. 3 ff.

2. D. I. Mackay and G. A. Mackay, *The Political Economy of North Sea Oil* (Martin Robertson, London, 1975), p. 33.

3. *North Sea Costs Escalation Study* (Her Majesty's Stationery Office/Department of Energy, London, 1977), pp. 8 ff.

4. Christopher Tugendhat and Adrian Hamilton, *Oil – the Biggest Business* (Eyre Methuen, London, 1975), p. 42.

5. Jens Evensen, *Oversikt over oljepolitiske spørsmål* (Industridepartementet, Oslo, 1971), p. 12.

6. Michael Tanzer, *The Political Economy of International Oil and the Underdeveloped Countries* (Temple Smith, London, 1969), pp. 59 ff.

7. Edith Penrose, 'Aspects of Consumer/Producer Relationships in the Oil Industry' in Ragaei El Mallakh and Carl McGuire (eds.), *U.S. and World Energy Resources* (ICEED, Boulder, Col., 1977), pp. 21–9.

8. Evensen, *Oversikt*, p. 10.

9. Guy Arnold, *Britain's Oil* (Hamish Hamilton, London, 1978), p. 92.

10. Sigmund Gjesdahl, 'Er vi blå-øyde arabere?' in *Dagbladet*, 3 September, 1978, pp. 3–4.

11. Brian C. Smith, 'Reform and Change in British Central Administration' in W. J. Stankiewicz (ed.), *British Government in an Era of Reform* (Collier Macmillan, London, 1976), pp. 215–33.

12. Helge Ole Bergesen, 'Uttappingstempoet – oljepolitikkens nøkkelbegrep' in Kari Bruun Wyller and Thomas Chr. Wyller (eds.), *Norsk oljepolitikk* (Gyldendal, Oslo, 1975), pp. 51–71.

13. Evensen, *Oversikt*, pp. 64 ff.

14. Kenneth W. Dam, *Oil Resources* (University of Chicago Press, Chicago, 1977), pp. 6 ff.

15. Hartmut Elsenhans, 'Entwicklungstendenzen der Welterdölindustrie' in Hartmut Elsenhans (ed.), *Erdöl für Europa* (Hoffmann und Campe, Hamburg, 1974), pp. 7–47.

16. Franklin Tugwell, *The Politics of Oil in Venezuela* (Stanford University Press, Stanford, California, 1975), p. 9 ff.

17. *Kostnadsanalysen – Norsk kontinentalsokkel* (Ministry of Oil and Energy, Oslo, 1980), Vol. 1, p. 22 ff.

2 NORTH SEA HISTORY

From the end of the Second World War until the 1970s Western Europe has become increasingly dependent upon imported oil. This was due to relatively rapid economic growth coupled with a relatively high level of energy consumption. During this period more energy-intensive patterns of consumption became widespread in Western Europe. Automobile ownership became common and fuel heating systems were installed in homes, offices and factories (see Table 2.1). Western Europe's coal reserves were mostly of poor quality and production was difficult to expand. However, this dependence on foreign oil was not seen as a problem because increasing quantities of oil were available in the world market at relatively decreasing prices.

Table 2.1: Western Europe: Consumption and Production of Energy, 1950–65 (million tonnes of coal equivalent)

Year	Total Energy Consumption	Coal Production	Oil Production	Oil Imports	Oil Imports / Energy Cons.
1950	584	488	6	57	9.8
1955	748	532	14	112	15.0
1960	850	500	23	191	22.5
1965	1,117	483	32	364	32.6

Source: Joel Darmstadter *et al., Energy in the World Economy* (Johns Hopkins University Press, Baltimore, 1971).

This increasing dependence upon imported oil gave all Western European governments a consumer interest in increasing oil supplies and low prices. As a result, Western European governments sought to accommodate the international oil companies, which at the time seemed the best guarantee for an uninterrupted flow of cheap oil. The UK and Norway were no exception to this rule, even if their degree of energy self-sufficiency was higher than most other Western European countries (Table 2.2). Coal in the UK and hydro-electricity in Norway meant a relatively moderate dependence upon oil imports.

Nevertheless, both countries were linked to the international oil industry in a more organic way. The UK harboured two of the world's

37

Table 2.2: UK and Norway: Consumption and Production of Energy
(million tonnes of coal equivalent)

	Energy Consumption	Energy Production	Oil Imports	Oil Imports / Energy Cons.
UK, 1950	223.1	220.2	18.2	8.2
UK, 1965	290.3	191.5	74.1	25.6
Norway, 1950	6.4	2.8	1.1	17.2
Norway, 1965	13.5	6.7	4.7	34.8

Source: Darmstadter, *Energy in the World Economy.*

leading international oil companies: British Petroleum was entirely
British-owned, Shell was 40 per cent British-owned, and both companies
had their headquarters in London. Further complicating the links, the
UK government had a 49 per cent interest in British Petroleum.

In Norway the merchant marine was heavily engaged in the tanker
trade, transporting oil for the international oil companies. The foreign
exchange earned in this way had a considerable importance for the
Norwegian current account balance. Consequently, in spite of a
relatively low dependence upon imported oil, both UK and Norwegian
government interests were to a considerable extent similar to or at least
intertwined with those of the international oil companies.

The international oil industry probably first took an interest in UK
and Norwegian waters after the discovery of natural gas in the
Netherlands in 1959. The new field was very large and stimulated
interest in the geological structures stretching north and west of the
Netherlands beneath the North Sea. The first offshore drilling took
place in Dutch coastal waters in 1961.[1] In 1962 the UK and Norwegian
governments received the first applications for permission to explore
for oil and gas on their continental shelves.

In 1962 the UK and Norway were not prepared to let private
companies explore for oil and gas. The continental shelf had not been
divided, and neither country had any legislation regulating economic
activities there. Although several countries had declared their
sovereignty over large maritime areas since 1945, those adjacent to the
North Sea had preferred to postpone such measures until there was an
internationally recognised legal framework.[2] The First United Nations
Conference on the Law of the Sea took place in Geneva in 1958,
adopting the Convention on the Continental Shelf. This convention
later gained wide international acceptance,[3] but by 1962 it had not yet

been ratified by the UK or Norway. Consequently, the UK and Norwegian governments were unable to respond positively to the applications. However, in both countries small groups of high-level civil servants started to work with the problems.

In the spring of 1963, slightly more than six months after the initial contacts with the oil industry, the Norwegian government declared its sovereignty over the continental shelf, explicitly making exploration for and the exploitation of natural resources a matter of national jurisdiction. The area under national sovereignty was defined as the sea bed adjacent to the coast of Norway where the depth of waters permits exploitation of submarine resources, but not beyond the median line relative to other countries.[4] A few weeks later, on 21 June 1963, a law was passed allowing seismic studies of potential petroleum resources, but not permitting drilling. The new law also delegated broad powers to the administration. The law established the principle that the rights to submarine natural resources belong to the state and that the right to explore for and exploit these natural resources can be granted only by the King, i.e. the government. Finally, in the Royal Decree of 9 April 1965 rules were laid down for exploratory drilling and exploitation of petroleum resources under the North Sea.

In the UK the Continental Shelf Act was passed on 15 April 1964. It declared the sovereignty of the UK over the continental shelf. Concerning the exploration and exploitation of petroleum resources, the Act explicitly referred to the Petroleum (Production) Act of 1934, which defined rules for granting licences for the exploration and production of petroleum in the United Kingdom. Unlike Norway, the UK already had legislation on the subject and this was applied on the continental shelf with a few modifications.

The physical characteristics of the North Sea created potential for conflict between the UK and Norway. Most of the bottom of the North Sea is separated from Norway by a rift, the Norwegian Trench, which drops to 700 metres (2,300 ft). On the UK side there is no similar feature. The depth of the trench at present does not permit exploitation of submarine resources and, if the Convention on the Continental Shelf were interpreted very literally, might have caused differences in interpretation. Disagreement over the partition of the shelf would have made it less attractive to the oil industry, and perhaps postponed exploration and production for many years.

Due to its relatively comfortable economic situation, Norway had no pressing need to start oil production. It could afford to postpone operations if there was a legal dispute. The UK was in a much less comfortable economic situation and was in a hurry to get exploration and production started. Consequently, the UK wanted to avoid any political tensions in the area, and was therefore happy to accept the Norwegian proposal of the median line.[5] On 10 March 1965 an agreement was reached dividing the continental shelf between the UK and Norway. The median line was measured from the UK coast and from the outer points and islands off the Norwegian coast, which was rather advantageous to Norway.

On a more rigid interpretation of the Convention, the Norwegian Trench might have been considered the outer limit of Norway's continental shelf. The parts of the shelf that are considered promising from the point of view of petroleum geology are essentially located to the west and to the south of the Trench, and could under other circumstances have fallen under the jurisdiction of the UK and Denmark.[6] Implications would have been profound for the economies of the three countries and probably for the rate of development as well. The ease with which the solution of the median line was agreed upon makes it legitimate to raise the question as to what extent the UK and Danish administrations realised the petroleum potential concerned, and to what extent they feared that Norway might retaliate, creating tensions that could retard the development.

In other parts of the North Sea the principle of the median line was not so readily accepted. West Germany contested Danish and Dutch boundary proposals. After prolonged negotiations the case was submitted to the International Court of Justice in the Hague in February 1967. The verdict of the court was ambiguous.[7] New bilateral negotiations followed, and only in November 1972 were agreements ratified.

The final partition of the North Sea gave the largest shares to the UK and Norway, smaller shares to Denmark, the Netherlands and West Germany, and minute fractions to France and Belgium (see Table 2.3). The UK got about 46.7 per cent, Norway about 25.1 per cent, the Netherlands about 10.7 per cent and Denmark about 9.2 per cent of the North Sea territory. In relation to land areas, the annexation of North Sea areas implied that the UK, the Netherlands and Denmark each more than doubled the territories under national jurisdiction and potentially open to economic activity. For Norway the annexation of the North Sea area implied a territorial increase of 'only' 41 per cent. However, in relation to population, Norway got by far the best deal, with an estimated 13,100 square miles of new territory per million

Table 2.3: North Sea Territories (square miles)

Country	North Sea Area	Land Area	Per cent Addition	Population in 1970 (thousands)	North Sea Area / Population
UK[a]	95,300	94,200	117	55,500	1.72
Norway	51,200	125,100	41	3,900	13.13
Netherlands	21,800	15,800	138	13,000	1.68
Denmark	18,800	16,600	113	4,900	3.84
West Germany	13,900	96,100	14	60,700	0.23

Note: a. Figures for the UK include parts of the continental shelf off the west coast.

Sources: D. I. Mackay and G. A. Mackay, *The Political Economy of North Sea Oil* (Martin Robertson, London, 1975), p. 21; and *Statistical Yearbook of Norway* (Central Bureau of Statistics), pp. 382 f.

inhabitants; for Denmark the figure was 3,800 square miles per million; and for the UK and the Netherlands the figure was around 1,700 square miles per million inhabitants. This difference in dimension implied different policies in relation to the new areas under national jurisdiction.

Along with the task of territorial partition, the governments faced a number of other complex questions before activities could start in the new areas. The basic problem for both the British and Norwegians was that the new areas were unknown from a petroleum point of view, and neither country had the appropriate experience and technology to find out about the new areas. Consequently, both governments had to rely on private and essentially foreign companies. This meant that the new areas had to be made attractive to the international oil industry, but at the same time both governments were under considerable political pressure at home to secure a fair share of the economic surplus for themselves, and in particular to stimulate the development of national expertise in the offshore petroleum industry.

Both governments realised that eventual discoveries of petroleum would make their continental shelves more attractive to the oil industry and consequently strengthen their bargaining position.[8] But they both faced the risk that international oil companies with access to several sources of oil, in particular to low-cost oil from the Middle East, might prefer to defer production of high-cost North Sea oil. Because of a more favourable economic situation and a relatively small population, Norway could afford to proceed more slowly than the UK. Because of the UK's economic troubles, it had a particularly strong need to accelerate the production of indigenous oil and gas resources.[9]

The UK trade-off was further complicated by one consideration that was irrelevant to Norway. It was believed that the UK, because of its size, would remain a substantial net importer of oil for a long time. It was explicitly feared that if the UK imposed stiff financial terms on the production of its relatively high-cost North Sea oil, it might cause OPEC countries to follow suit.[10] This could in turn hurt the UK balance of payments and the operations of UK oil companies in OPEC countries.

In elaborating the regimes for opening up new areas, the UK and Norwegian governments had to solve some important questions related to their strategies as landowners and the practical functioning of the oil industry in the new areas. These questions concerned the method of allocation, the size and duration of concessions, royalties and duties, and the organisational pattern.

From the outset it was clear that both governments saw their interests in oil policy as much more than purely financial. In fact, obtaining revenue from oil activities in the North Sea was only one of several preoccupations, and it was hardly the dominant one at all times.

It is this diversity of government interests that is the clue to understanding the complexity of UK and Norwegian oil policy. It certainly explains why both governments have quite systematically stuck to the method of discretionary allocation and avoided the auction method. The auction method of allocation is thought to give governments the highest possible proportion of economic rent.[11] However, in the context of the North Sea oil industry, both the UK and Norwegian governments have generally felt that they could gain more control through discretionary allocation.[12]

By 1964–5 both the UK and Norway had formulated their first comprehensive regimes for the exploration and production of petroleum in the North Sea. The two initial regimes had some important features in common. Both governments had the right to control the working programme, and both countries from the outset opted for a system of relinquishment, which forced companies to hand back part of their concessionary areas after a given period of time. Both countries opted for a relatively mild form of taxation.

The Beginning

Following the passing of the Continental Shelf Act in the spring of 1964 the UK Conservative government invited applications for licences. Applicants had to be incorporated in the UK so that their profits would be taxable there. The criteria for allocation were based on the proposed working programmes, the abilities and resources to implement them and their contributions to domestic energy resources in general. Foreign applicants were to be treated as UK companies were treated in their country of origin.[13] The result was that 53 licences were granted, mainly to foreign companies. UK interests represented about 23 per cent, and public sector involvement was about 9 per cent. There was some sign of preferential treatment in that UK companies were favoured in the more promising areas.[14] Gas in commercially exploitable quantities was found in the southern part of the UK sector in 1965.

In Norway the Labour government invited applications on rather similar terms. Altogether 78 licences were granted in the spring of 1965 to nine groups of companies, with American and French interests being prominent. Norwegian interests were represented through Norsk Hydro, Scandinavia's largest chemical firm, which is partly government-owned. In the Norwegian area exploration was less successful. Gas condensate was found in 1968, but in quantities not judged commercially exploitable.

In the UK, after the elections of October 1964 the Labour Party took over the government. The new administration reviewed oil policy and new criteria for licence allocation were defined. Experience, a contribution towards the UK balance of payments, the growth of public and private industry, employment and regional development were added. The second round of licensing took place in the summer of 1965. Thirty-three new licences were granted, with UK interests representing about 34 per cent, and public sector involvement representing about 16 per cent.

In Norway the Labour government was defeated in the elections of September 1965, and a liberal–conservative coalition took over. Oil policy was left basically unchanged, as was the administration of matters relating to oil. The criteria for licence allocation were modified, with provisions for state participation being included. Applications for new licences were invited in May 1968. In 1969, 14 new licences were granted to six groups of companies, all of which were already present in Norwegian waters. All the new licences provided for some form of state participation in the form of net profit sharing and carried interest.

In the UK sector significant discoveries of natural gas were made in 1966 and 1969. This strengthened the bargaining position of the British and caused them to review both their licensing policy and their energy policy. In particular, the question of state participation received a lot of attention. A proposal from a Labour Party study group to create a national hydrocarbon corporation, a wholly state-owned oil company, was discarded. Instead it was decided to let the state participate through the Gas Council and the National Coal Board. In September 1969, 37 new licences were granted in the third UK round, and this time UK interests represented about 37 per cent, with the public sector holding about 20 per cent.

In the Norwegian area oil was first struck in December 1969. In the summer of 1970 it became evident that the oil discovery in the Ekofisk field was not only commercially exploitable, but of considerable proportions. The oil discovery had an important impact upon Norwegian oil policy. The liberal–conservative coalition government decided to increase its participation in Norsk Hydro from 48 to 51 per cent. However, at this time the Norwegian political system became absorbed in the Common Market issue, which overshadowed oil policy. In March 1971 the liberal–conservative coalition government, torn by dissent on the Common Market issue, was replaced by a minority Labour government. The new government decided to review oil policy with an emphasis on state participation and a more restrictive approach. In June 1971 a single licence was issued to a group of American and Norwegian companies, with the state reserving the right to participate at up to 26 per cent. In September 1972 the Norwegian Parliament voted unanimously to establish a wholly state-owned oil company, Statoil, to take care of the state's commercial interests on the continental shelf as a wholly integrated oil company.

After the referendum on the Common Market question in September 1972 the Labour government withdrew and a minority liberal-centrist government was established, which was made up of parties that had opposed entry into the Common Market. This government granted only one licence in September 1973, which provided for state participation of 50 per cent through Statoil. Invitations were issued for applications for a considerable number of licences in the summer of 1973. But it was decided that, before licences could be granted, the entire Norwegian oil policy should be reviewed by Parliament. With the Common Market issue closed, oil policy could again get the full attention of the political system. This was an urgent matter because, in the wake of the oil discovery and partly as a function of government neglect, a large number of small oil companies had mushroomed and speculation in oil shares and in rights for oil shares was rampant. Perhaps up to 10 per cent of Norway's adult population possessed either oil shares or rights for such shares. Most of these newly created companies had extremely limited technical and financial resources and were later judged unfit to take part in oil activities on the shelf.[15] Instead, in 1972, the Norwegian Ministry of Industry took the initiative to co-ordinate and concentrate private industrial interests desiring to take part in the oil activities into a new company, Saga Petroleum.

In the UK the Conservatives took office after the elections of June 1970. At this time discoveries in both the UK and the Norwegian sectors of the North Sea indicated considerable petroleum potential and it was thought that the northern part was particularly promising. This element seemed to strengthen the bargaining position of the UK government, but factors in the international oil market worked the opposite way. In 1969–70 Libya was successful in increasing its revenue from oil and other OPEC countries were following suit. The result was an additional burden on the UK's balance of payments in the form of more expensive oil imports.[16] This prompted the UK government to move ahead more quickly with domestic oil production and to review the licensing system.

The aims of British oil policy shifted to a maximum effort at exploration and development, combined with a good representation of British interests.[17] On the whole, the system of discretionary allocation was believed to provide the government with satisfactory control of the activity. It was in particular considered important in order to give preferential treatment to UK companies and stimulate the development of national expertise. Control over the working programmes of each particular area also guaranteed thorough explorations.[18] On the other hand, the system of discretionary allocation did not give the government the best financial results, and some felt there should be a more conscious balancing of development and financial concerns. This resulted in the decision to retain the system of discretionary allocation, but to experiment with the auction system.

In June 1971 the UK government invited applications for the fourth round of licensing. Applicants for 15 areas to be auctioned were invited to submit tenders for initial payments. Licences for the 15 areas were granted to the highest bidders, and the sum of the highest bids for each area totalled £37 million. Another 118 licences were granted by discretionary allocation. The results of the two methods show striking differences. In the 15 areas auctioned, UK participation totalled 20 per cent and public sector participation came to 10 per cent. In the areas subject to discretionary allocation, UK interests represented about 35 per cent and public sector participation was at about 10 per cent. The auction system seemed to favour large foreign oil companies to the detriment of UK ones, and, if applied generally, could compromise the aim of developing national expertise and technology. However, the results also showed that the auction system could be much more financially beneficial to the government than the discretionary system, as the payment of normal licence fees totalled only £3 million.[19]

In the early 1970s there was widespread opinion in the UK that the Government did not get a fair share of the revenues from oil. British corporations had free depreciation, meaning in practice that any capital costs could be written off within one single year. For oil companies this included exploration costs. Also, taxes paid on foreign operations were set against UK taxes and not against the corporate profits. In addition, companies that were members of a group operating in the North Sea could get relief from income taxes by setting profits against group losses, and vice versa. Finally, when tax deductions in one year exceeded UK income taxes due, they could be carried forward to offset tax obligations in coming years. It was estimated that nine major oil companies had in this way accumulated 'tax losses' of £1,500 million by 1972.[20] It was feared that 'tax losses' could be accumulated indefinitely and thus effectively deprive the UK government of any corporate profit tax from the major oil companies operating in the North Sea. Between 1965 and 1973 the major oil companies paid only £500,000 in corporate taxes.[21] The problem was further complicated by the fact that many of the same companies were producing oil in OPEC countries, importing and selling this oil in the UK and at the same time developing oil production in the UK area of the North Sea. The OPEC posted price, which was a tax reference price and in excess of the real sales price, was often used as a transfer price within oil companies. Thus, these companies could develop profit centres abroad not liable to the UK corporate profits tax, at the same time increasing their 'tax losses' in the UK.

At this time the production of oil in the UK sector of the North Sea had hardly begun, but, if permitted to continue under the existing taxation system, it would probably have given the government returns from the corporate income tax that over many years would have been embarrassingly close to zero.[22] Correspondingly, it has been calculated that the initial system would have given extremely high rates of return to the companies, in some cases close to or exceeding 100 per cent.[23] This was much more than could reasonably be judged necessary to attract risk capital to North Sea operations. The average rate of return on capital invested in UK industry in the 1970s appears to be around 15 per cent.[24] Consequently, it was recommended that action be taken to improve the government tax yield, and that no new licences be granted until all aspects of UK oil policy, including licensing and taxation, had been reviewed.[25]

By the early 1970s oil had become an integral part of UK and Norwegian political life. In the UK licensing methods and taxation were the great issues, in Norway the role of private companies and the rate of development were dominant questions. In both countries the Labour parties favoured a stronger degree of state participation, and even the non-socialist parties seemed to agree that changes ought to be made in taxation and in the organisation of the oil industry. Both countries had acquired some valuable experience, and since the early 1960s the terms worked out between governments and companies had gradually changed to the benefit of the governments. Important factors in this process were:

(1) the certainty that there were considerable quantities of oil in the North Sea;
(2) the increasing competence and insight of governments in matters relating to oil; and
(3) the gradual change in the world oil market in the early 1970s, which was epitomised by higher OPEC prices and participation agreements.

The changing circumstances in the world oil market made the need for a revised government oil strategy in the UK and Norway more acute. Consequently, UK and Norwegian oil policies were in a state of flux in the summer of 1973.

A New Situation

The changes in the international oil market in the autumn of 1973 completely altered the basis of oil activities in the North Sea. Clearly the economics of North Sea oil were affected by the fourfold increase in the price of oil. Politically, the bargaining situation between governments and companies was profoundly modified by the prospect of oil shortages and by the increasingly strong tendency of OPEC countries to nationalise their oil production. Given these changes, North Sea oil appeared much more attractive from an economic point of view, and more necessary from the point of view of supply security.

But these changes also strengthened the position of governments in relation to companies. Just as the oil companies could get windfall profits on their North Sea operations from the OPEC price rise, the actions of the OPEC countries also gave the UK and Norwegian governments significant political benefits. Without being affiliated with OPEC, a community of interests developed between the North Sea oil producing states and OPEC.[26] This in turn has affected the oil policies of both the UK and Norway.[27]

At the time of the oil crisis the UK and Norwegian governments were actively reviewing key aspects of their oil policies. In the UK the Conservative government had declared that it was dealing with the tax issues raised by the Report of the Committee on Public Accounts.[28] A new taxation system, including an excess profits tax of 75 per cent, seems to have been in preparation[29] in the first few months of 1974.

In Norway the liberal-centrist coalition government gave oil policy top priority once the negotiations with the Common Market on a trade agreement were concluded.

Even in the winter of 1972–3 the question of landing oil and gas from the Ekofisk field caused a major political controversy in Norway. Parliament decided that in principle oil and gas should be landed in Norway. However, the existence of the Norwegian Trench, separating the major oilfields from the coast, made the construction of pipelines to Norway out of the question. Furthermore, most of the oil and gas would have to be re-exported from Norway. In the spring of 1973 it was decided to build a pipeline for oil to Tees-side in the UK and one for gas to Emden in West Germany.[30]

This decision caused much opposition because it was feared that Norway's control over its oil was lost. In the early summer of 1973 the government commissioned the administration to prepare two reports on oil policy, one on the organisation of the activities on the continental shelf and another on the role of petroleum industry in Norwegian society, dealing with macro-economic and social aspects.

In the electoral campaign of 1973 oil policy became an important issue. The Labour Party declared it would curtail the growth of private speculative companies, increase state participation and opt for a moderate rate of development in order to have balanced economic and social development.

After the elections of September 1973 the Labour Party took office again, backed by a relatively strong left-socialist representation in Parliament. The work on the reports initiated by the preceding government was continued, but with new directions. During this time the premises of the work were changed dramatically with the development of the oil crisis. The reports were submitted by the government to Parliament and published in the spring of 1974. They merit attention, as they give a comprehensive picture of the oil policies proposed, and they have subsequently had considerable impact.

The Parliamentary Report on the Petroleum Industry in Norwegian Society[31] recommended a restrictive approach to oil, emphasising the need for a moderate rate of development and the need for public control over all important aspects of the petroleum industry. A moderate rate of development was seen as a level of production of approximately 90 million tonnes of oil equivalents per year. Also, moderate use of incomes from petroleum was recommended in order to control inflationary pressures and changes in the patterns of production and settlement. In order to exercise control an increasing and progressive system of state participation, through Statoil, was recommended. It should be underlined that the Report did not recommend full nationalisation of the oil industry, and it explicitly assigned a role for private and foreign companies. Nevertheless, a revision of taxation policy was announced.

The report discussed at length the impact of the oil industry on the Norwegian economy. As there was already virtually full employment in Norway, considerations of economic and social policy as well as of resource management were arguments in favour of a cautious approach to oil. Despite this caution, the need for Norwegian expertise and technology was also stressed. With regard to foreign oil policy the report announced the intention to develop closer contacts and co-operation with OPEC countries. The report also announced the government's intentions in economic policy, particularly concerning the use of oil revenues. Tax deductions and increased efforts in social, regional, educational and environmental policy got a high priority, the goal being to make Norway 'a qualitatively better society'. In order to ensure better public control of industry, it was proposed that part of the oil revenues be used to buy up private interests in Norwegian industry, including foreign interests.

The report from the Ministry of Industry on the Operations on the Continental Shelf[32] was of a more technical nature, but it essentially followed the same lines as the preceding report. It recommended state participation as a tool of control and as a means of securing revenues for the government. The system of carried interest, already practised, was considered particularly useful. With regard to licensing policy, the use of the auction system was not even mentioned. The report stressed the need to develop Statoil as quickly as possible, so that it could acquire the expertise required to be the practical and commercial instrument of Norwegian oil policy.

In the longer run it was recommended that the government, through Statoil, should carry the risks of exploration and development alone. But the need to co-operate with other companies was also stressed because of the need for risk capital and the lack of Norwegian experience. It was explicitly stated by the Ministry of Industry that a new area could be explored most successfully if several different companies were admitted, with different geological theories and levels of experience. Finally, the report explicitly excluded other Norwegian companies such as Norsk Hydro and Saga Petroleum from the continental shelf because of their inadequate technical and financial resources.

The announced oil policy was opposed from several quarters. Interests that might have profited from a higher rate of development and a greater role for private enterprise felt cheated. This included private banks, certain private oil interests and the private shipowners who had invested heavily in drilling rigs. In one journal it was argued that the exclusion of private Norwegian oil companies from the continental shelf was a logical result of the socialist–communist philosophy, and that private industry must fight the socialist–communist parties.[33] The Norwegian Shipowners' Association accused the government of unduly politicising the oil issue by using the oil revenues to turn Norway towards socialism and undermining confidence in private industry.[34] However, the Norwegian Federation of Industries had a more cautious and balanced approach to the new oil policy, even if it did not approve of using oil revenues to buy up industrial interests. Quite separately the government was subject to criticism from a wide variety of interests fearing a rapid pace of oil development. These interests included fishing, agriculture, many smaller enterprises, labour-intensive industries fearing a higher cost pressure, and many wage-earners fearing inflation and a more unequal distribution of income. Here there was enthusiasm for keeping oil under firm public control, but the level of production announced was thought by many to be too high.

The debate in Parliament in the early summer of 1974 showed that the government at least politically had struck a fair balance between opposing groups. In fact the proposed policy was approved by large parts of the opposition as well, which created a general consensus on the principles of oil policy in Norway. In accordance with the new policy, a much smaller number of licences were granted in the autumn of 1974. This time Statoil was to have carried interests of 50 to 55 per

cent in each licence with the option of increasing participation up to 75
per cent, depending on the size of the eventual discoveries.[35]

In the UK the miners' strike at the time of the oil crisis provoked a
national emergency, a three-day working week, and finally elections in
February 1974 that put Labour in power. Oil policy was one of the
themes of the campaign. The Report of the Committee on Public
Accounts, combined with the rise in oil prices, had added heat to the
controversy. In its election manifesto, Labour had announced its
intention of ensuring full public ownership of petroleum resources and
majority public participation in oil operations.

In July 1974 the new government published a White Paper entitled
United Kingdom Oil and Gas Policy.[36] The White Paper emphasised the
potential losses to the country under existing arrangements. Oil
production was projected to reach 100–150 million tonnes a year in
the 1980s. At constant oil prices pre-tax profits might total £4,000
million by 1980 with taxes and royalties only taking a relatively small
proportion. If at that time only a part of the oil company profits were
transferred abroad, the resulting burden on the balance of payments
would be intolerable. Consequently the government intended to increase
the government take and assert greater public control. The White Paper
proposed the creation of a special additional tax on oil companies
working in the North Sea in order to prevent losses of revenue. In
addition, a system of state participation in new licences was announced
that resembled the pattern of carried interest practised in Norway. The
government also intended to renegotiate existing licences to obtain
majority state participation throughout.

Furthermore, the White Paper announced the creation of a wholly
state-owned oil company, the British National Oil Corporation, BNOC,
which would exercise the state participation rights. The government
also planned to extend its powers to control the physical aspects of oil
production and pipelines in order to assure environmental protection
and the proper planning of infrastructures.

The proposed changes in taxation were quite radical and included
the cancellation of artificial losses accumulated before 1 January 1973,
the introduction of a transfer price to be used in tax computations, and
the introduction of a so-called 'ring-fence' in order to calculate the
profits and losses of North Sea operations separately from other
company activities. Finally, the White Paper made special mention of
the need to let Scotland benefit from the development of North Sea oil.

The proposals of the White Paper met with surprisingly little
opposition in the UK,[37] mainly because of the widespread impression

that the oil companies were reaping extraordinary profits at the expense of the Treasury and the public. Rising Scottish nationalism also made the British public more sensitive to the need to change oil policy. Finally, the example of Norway was also important. In the early 1970s the British public had the general impression that Norway 'had husbanded its North Sea oil potential with far greater skill and fore-sight'.[38] In addition, the oil policies of the UK Labour Party were certainly influenced by the thinking and practices of Norwegian Labour governments.[39]

In Norway, preparations were also made to change the oil taxation system. The explicit intention was to secure a fair share of the windfall profits generated by the extraction of oil. Initially there was a dilemma over whether to renegotiate, because of the radically altered circum-stances in the world oil market, or whether to solve the problem through new taxes. The dilemma essentially concerned the Ekofisk field, where development was at an advanced stage by 1974 and enormous profits were possible for the companies involved. Imposing state participation might have created difficulties in international law, so instead it was decided to propose a 40 per cent tax on excess profits, thus assuring a total state share of profits of 90.8 per cent. This proposal was rejected by the companies involved in the Norwegian sector of the North Sea, and the government had to back down in December 1974. A new proposal presented in February 1975 provided for a special excess profits tax of 25 per cent and included many possible tax deductions. This provided the government with a maximum share of profits of 75 per cent and an average share of between 57 and 66 per cent. This amended proposal was accepted by the companies and was presented to Parliament in the spring of 1975.

In the UK an Oil Taxation Bill was prepared in the autumn of 1974 and became law in the spring of 1975. The law introduced a petroleum revenue tax on certain assessable profits, with special provisions for tax deductions that were to be calculated individually for each field.[40]

In 1975 the British National Oil Corporation was established and negotiations began with the companies already operating in the North Sea to establish majority BNOC participation. The first participation agreement was concluded in January 1976, between BNOC and Gulf/ Conoco. The UK form of participation in existing licences differs considerably from the Norwegian pattern. BNOC usually got the option to buy up to 51 per cent of the oil at market prices, which is a form of participation that might be called a purchase agreement.[41]

Thus, by 1975/6 both the UK and Norway had new oil regimes that implied a profoundly modified relationship between governments and companies. The two regimes are basically similar even if their technical and organisational solutions differ occasionally. The most obvious common features are state participation and special taxation, which reflect the governmental aims of exercising more control and securing a larger share of the economic rent. However, as noted above, the participation and taxation practices are different in the two countries. This reflects different needs and priorities and perhaps differences in bargaining positions. In Norway there is a definite continuity in oil policy, with gradually tougher terms being imposed by the government. In the UK, by contrast, there is a marked discontinuity. A very liberal regime was replaced within a short period of time by a much more restrictive one.

Both governments suffered certain setbacks in their bargaining with the companies, at least in relation to their original intentions. The UK government had to yield on the question of participation in existing licences, the Norwegians had to yield on the tax question. However, in a short time both governments had very considerably improved their positions in relation to the private oil industry. This was owing to a combination of circumstances: improved insight on the part of governments, public pressure and, of course, the oil crisis. Although the oil crisis had a major effect upon the bargaining situation, it is reasonable to assume that important changes would have taken place in UK and Norwegian oil policies in any case, given the domestic political pressures.

The greater continuity in Norwegian oil policy can partly be explained by the overall economic situation that permitted a more restrictive policy at an earlier stage of development, but was also due to Norway's stronger tradition of economic planning. The Norwegian Finance Ministry, which co-ordinates economic policy, seems to have more varied functions and stronger powers than the UK Treasury, which is more confined to public finance and budget matters. The fact that the UK has a formal economic planning body, the National Economic Development Council (NEDC), and Norway does not, seems to have been of little relevance. In the UK the Department of Trade and Industry (DTI) appears to have been more dominant than the Norwegian Ministry of Industry. This probably meant that foreign considerations played a more important role in the early stages of UK oil policy, namely the defence of UK oil interests abroad and the protection of the interests of the international, mostly American, oil companies. These issues played little part in the elaboration of

Norwegian policy, partly because the Norwegian Finance Ministry and Ministry of Industry have traditionally been concerned with domestic matters, and partly because the Norwegian Ministry of Trade and Ministry of Foreign Affairs seem to have been excluded from some of the critical decision-making. Also, the recruitment to important bodies in the two countries seems to be different. In Norway the staff dealing with oil policy in key Ministries tended to be young economists specialising in oil matters, whereas in the UK the older civil servant with a more generalist approach has perhaps been more important. Thus, the difference in performance could also be explained by differences in bureaucratic structures and recruitment. However, it should be noted that although Norway asserted 'oil nationalism' earlier, this approach has now largely been adopted by the UK. 'Oil nationalism' is in keeping with the traditions of government in the UK and Norway. Both governments had to respond to public demands that the oil industry be controlled and perform according to national needs; an unfettered oil industry would probably have compromised important national interests. In the UK the repatriation of huge profits could have been a new burden on the balance of payments. In Norway, large sectors of the economy would have been threatened by uncontrolled expansion of the oil industry and the relatively equal distribution of income, as well as the regional balance of the country's economy, would have been compromised. Norway's early rigour in oil policy was much more a measure of self-defence by an integrated and relatively homogeneous society than a sign of excessive economic radicalism. If this approach worked in Norway, it could also work in the UK. Given these political realities, the oil industry should perhaps be glad that a *'laissez faire'* approach was not attempted in either country. This would most likely have created acute political conflict in relation to oil and maybe stronger demands for outright nationalisation. The fact that oil companies do not produce votes makes them vulnerable in critical political situations.[42] Therefore, it is in the interest of the oil industry itself to avoid serious dispute with the political environment in which it operates and to respect local political standards. This is particularly true of the more delicate Norwegian context, but it also applies to the scene in the UK. In this way long-term security could be offset against short-term excess profits.

There has been a marked difference in industrial performance between the UK and Norway in relation to the oil industry. As early as 1966 several Norwegian shipowners and shipyards decided to go into the offshore oil market. Before any significant amount of oil was found,

Norwegian shipowners ordered drilling rigs, service vessels, etc., from local yards. Norwegian rig designs were developed with such success that the American monopoly was broken, and by the mid-1970s rigs of Norwegian design had a strong position in the world market. In addition, Norwegian industry has been a pioneer in the development of concrete production platforms. A consortium of local constructors, Norwegian Constructors, pooled resources and developed the integrated platform, Condeep. By contrast, in the UK by 1975 there was only one yard building drilling rigs, there was scarcely one UK rig design, and few rigs are wholly UK owned. There have been considerable delays in the building of production platforms, and most platforms built in the UK are of foreign design. The UK record for supply ships is also poor. More recently, particularly with equipment and installations on the platforms, the British have performed at the Norwegian level.

Norwegian industry has not been able to follow up the success of Condeep. Many small firms are competing and specialising in the same fields and seem to be missing orders. Some larger Norwegian firms have co-operated with foreign partners in order to gain experience and know-how, but the results have not been very satisfactory.[43] Partly because of increased government efforts in recent years, UK industry has been able to step up its deliveries of oil drilling equipment and to get a larger share of the market.

How can we account for these differences in UK and Norwegian industrial performance? In the 1960s and early 1970s UK shipbuilding, steel and construction were working at or close to full capacity. They thus saw little need to develop new products for offshore oil development. UK industry and shipowners have a certain aversion to taking risks and venturing into completely new areas; shipyards were beset by notorious problems of management and labour relations in the 1960s and 1970s, and UK capital structure was probably also an important factor. The financial interests in the City of London ignored the potential market offered by the oil activities in the North Sea, or preferred higher or more secure profits elsewhere. Later, as the activities were developed and there was a need for new sophisticated types of equipment and installations, the large industrial base and innovative ability of UK industry were able to cope with the task.

Norwegian shipowners have a tradition of gambling and taking risks. They are accustomed to risking large sums for the sake of quick profits, which has produced an acute sensitivity to new opportunities. The gambling attitude, however, led them to over-invest in rigs and tankers in the wake of the oil crisis in 1973. As a result Norwegian shipping and shipyards are at present in a profound structural crisis because of over-capacity. The inability of Norwegian industry to follow up the Condeep success and to cope successfully with bids for equipment and installations is essentially due to its limited industrial base and the small size of firms. In general the Norwegian companies have inadequate resources, management that is usually oriented towards production rather than marketing, and problems of co-operation between the many small firms.

North Sea Oil in the Late 1970s

In August 1976 the UK government invited applications for the fourth round of licensing. Terms were that BNOC, alone or together with another state corporation or subsidiary, should have a minimum of 51 per cent participation, except in cases requiring an exclusive state licence. In August 1978, in the sixth round of licensing, private companies were invited to submit tenders for 46 blocks, including some in western waters. Terms were tougher than before. The private companies were asked to offer BNOC an equity stake higher than 51 per cent.[44] In August 1978 a revision of UK oil taxation increased the government take. To a large extent as a result of increasing oil production, the UK current balance showed a surplus in the first half of 1977, which permitted a more expansionary economic policy. In the spring of 1978 a White Paper on the future use of public oil revenue[45] gave high priority to industrial investment and improving industrial performance. The Conservative government elected in May 1979 essentially appears to retain the oil regime put in place by its predecessor.

In April 1976 the Norwegian government announced that drilling for oil exploration would start in the northern waters, north of 62° N, by 1978. Exploration was planned off central Norway and off the extreme north of the country.[46] Seismic surveys indicated geological structures that might contain large quantities of petroleum. In March 1977 the Norwegian government further announced that 15 new areas would be opened for exploration in the southern sector, south of 62°N. One of these areas was thought by some specialists to contain anything from 5 billion to 15 billion barrels of oil.[47] For the first time the government proposed risking state capital for exploratory drilling in this area.

In the spring of 1977 conditions were ripe for an increase in oil activities on the Norwegian continental shelf. A higher level of activity had become more of a necessity if the country was to maintain a steady increase in real wages and living standards in the middle of an international recession. However, this acceleration of oil development was met with criticism for environmental and safety reasons. In parliamentary debates it was remarked that there was a definite chance of a blowout occurring in the North Sea, and that protection measures were still inadequate. This criticism was discounted by the government.

A few weeks later, on 22 April 1977, a blow-out did occur, on the Bravo platform of the Ekofisk field. It was of rather small proportions and it was stopped in seven and a half days. Only limited quantities of oil were involved and the damage to the environment was insignificant.[48] Despite these objective facts the blow-out had a direct effect on Norwegian oil policy. Parliament decided to postpone the opening of new areas, and the government decided to defer exploratory drilling in the northern waters until 1980. It was generally felt that the blow-out should be analysed and measures relating to safety and environmental protection should be improved.

In the autumn of 1977 the government made a new proposal on exploratory drilling in the southern waters. In a political compromise the promising area mentioned earlier was reserved for Statoil, but with a somewhat higher degree of participation by Norsk Hydro and Saga Petroleum than was originally intended. For the other licences terms were tougher than in previous licensing rounds. A new clause entitled the government to hold back the development of oil discoveries for a specified number of years. The degree of state participation was to be a minimum of 50 per cent, escalating up to an unspecified ceiling. Also, licensees were to order equipment in Norway, and the government could use the creation of jobs and the access to industrial expertise outside the oil sector as a criterion for licensing.

In early 1978 negotiations on industrial co-operation were going on between Norway and several countries, particularly Sweden and West Germany. In the late spring the coupling of the Norwegian purchase of part of Swedish Volvo with Swedish access to Norwegian oil was announced. This deviated from the traditional Norwegian policy of keeping oil licensing isolated from other deals. Drilling in the southern sector was resumed in 1978, and in August 1978 it was announced that oil had been struck by Statoil in its reserved area. The blow-out, slower licensing and delays in the development of Statfjord combined to reduce the level of activity in the Norwegian oil sector.

In the meantime the Norwegian economy had gradually deteriorated. In the spring of 1978 Norway's foreign debts reached $20,000 million, corresponding to half of the country's GNP. This is the highest debt ratio ever reached by an OECD country. In 1977 and 1978 Norway had the highest current deficit in the OECD area, with the exception of the United States. About half of the foreign debt was due to investment in the oil sector[49] and the rest was caused by government borrowing that was meant to rescue insolvent Norwegian shipowners and to maintain a high level of economic activity in Norway. The decline in world shipping hit Norwegian shipowners hard. The recession in the Western economy cut demands in the traditional Norwegian export markets of Sweden, the UK and West Germany. Simultaneously higher domestic production costs caused Norwegian industry to become less competitive and traditional exports declined in volume.

The result of the government's anticyclical policy was to maintain a high level of employment, which caused real wages to increase by about 25 per cent between 1973 and 1977, a period of international industrial recession. Norwegian labour costs for industry are now possibly the world's highest.[50]

After the parliamentary elections in September 1977, a major reversal of economic policy was announced. In the spring of 1978 a revised version of the long-term programme was presented, announcing severe measures of austerity and a halt to real wage increases.[51] The combination of a large foreign debt, increasing debt service payments, reduced and deferred oil revenues, declining traditional exports and increasing imports of consumer goods would change the policy of any country. Furthermore, in Norway there have been explicit fears that the country might be put under economic administration by the International Monetary Fund unless policies were drastically changed.[52]

Against this background it appears that the Norwegian government to a certain extent has outsmarted itself. The cautious approach to oil seems most sensible in the Norwegian situation, and it is equally sensible to spend part of an anticipated financial surplus beforehand in order to maintain a high level of economic activity. However, the inability to control oil development and the vagaries of international economic performance caused these sensible policies to backfire.

With regard to oil, it is evident that the pattern of organisation has not given the Norwegian government full control of the trade-off between time targets and cost targets, nor adequate current information on development. With regard to the economic recession, it appears that the Norwegian government believed for several years that it would be only a short interruption in a continuous pattern of growth.

Norwegian oil policy also merits critical attention. A somewhat more

Table 2.4: Licensing Rounds, UK

Round	Year	Licences Granted	Companies	Area Granted (sq. miles)	British Interests (per cent)	British Public (per cent)	British Private (per cent)
1	1964	53	51	32,000	22.7	9.2	13.5
2	1965	37	44	10,000	33.6	15.5	18.1
3	1970	37	61	8,000	36.5	20.0	16.5
4 (a)	1971–2	118	213	24,000	20.0	10.0	10.0
4 (b)					34.7	9.6	25.1
5	1978–9	28	64				
6	1978–9	23	59				

Source: *First Report from the Committee on Public Accounts* (HMSO, London, 1973), p. 45 and subsequent annual reports by the Department of Energy.

(a) by auction, (b) by discretionary allocation.

vigorous approach to oil licensing in the early 1970s might have created a more favourable situation. Production and revenues would certainly have been higher, improving the balance of payments and the government's liquidity. Also, the Norwegian economy would have been less vulnerable. In Norwegian waters, the Statfjord field, which was found in 1974, is essentially the only large oilfield being developed in the later 1970s. Even if the field is large, with an estimated potential of about 3.5 billion barrels, its profitability for the oil industry and the government is finite. Finally, cost increases and delays have been more serious in Norwegian waters than on the UK side, largely because of the relative importance of one single field. These higher costs and delays in turn contribute to the low level of activity because they reduce the incentives for expansion. A higher level of development might have been an advantage to the Norwegians for two reasons. First, by developing more fields at the same time more experience would have been acquired. Second, by developing more fields there might have been more competition among furnishers of equipment and services.

At least part of the cost escalation probably consists of monopoly profits. Since 1973 development costs in the North Sea have risen much more than the general rate of inflation, and this additional price rise can hardly be explained by increasing technical complexity alone. In the Statfjord field, where the cost escalation has been particularly strong, only a few companies are furnishing goods and services, and one American agent in particular, Brown and Root, has a virtual monopoly in some key areas. This US company is consequently subject to increasing criticism in Norway.[53]

The Norwegian government will for several years be extremely dependent upon the smooth operation of two oilfields and one gas field for its foreign exchange. Even if the fields are of large proportions, this concentrated dependence upon a few oilfields makes the Norwegian economy extremely vulnerable.

By contrast, in the UK, where initial ambitions were smaller, the growth of production costs has been less pronounced and there have

Table 2.5: Licensing Rounds, Norway

Round	Year		
1	1965	5	1979–80
2	1969		
3	1978		
4	1978–9		

been fewer delays. Consequently, the UK economy is less dependent upon a few oilfields.

Table 2.6: Production of Oil and Gas, UK

Year	Oil (m tonnes)	Gas (m cu. metres)
1967	–	50
1968	–	2,050
1969	–	5,140
1970	–	11,270
1971	–	18,670
1972	–	27,180
1973	–	29,590
1974	–	35,730
1975	–	37,230
1976	12	39,410
1977	38	40,000
1978	58	38,000

Source: *Development of the Oil and Gas Resources of the United Kingdon 1977* (Department of Energy, London, 1977), pp. 3 and 37.

Table 2.7: Production of Oil and Gas, Norway

Year	Oil (m tonnes)	Gas (m cu. metres)
1971	0.3	–
1972	1.6	–
1973	1.6	–
1974	1.7	–
1975	9.2	–
1976	13.6	–
1977	13.4	2.5
1978	17.2	13.4

Source: *Fact Sheet: The Norwegian Continental Shelf* (Ministry of Industry, Oslo, 1977), p. 4.

Notes

1. Irvin L. White, Don E. Kash, Michael A. Charlock, Michael D. Devine and R. Leon Leonard, *North Sea Oil and Gas* (University of Oklahoma Press, Norman, Oklahoma, 1973), p. 3.
2. Patricia W. Birnie, 'The Legal Background to North Sea Oil and Gas Development' in Martin Saeter and Ian Smart (eds.), *The Political Implications of North Sea Oil and Gas* (Universitetsforlaget, Oslo, 1975), pp. 19–50.

3. Ibid., p. 19.

4. *Operations on the Norwegian Continental Shelf*, Report No. 30 to the Norwegian Storting (1973–4), (Ministry of Industry, Oslo, 1974), pp. 6 f.

5. Louis Turner, 'State and Commercial Interest in North Sea Oil and Gas: Conflict and Correspondence' in Saeter and Smart, *The Political Implications of North Sea Oil and Gas*, pp. 93–110.

6. Keith Chapman, *North Sea Oil and Gas* (David and Charles, London, 1976), p. 71.

7. Ibid., p. x.

8. Jens Evensen, *Oversikt over oljepolitiske spørsmål* (Ministry of Industry, Oslo, 1971), p. 13.

9. *First Report from the Committee on Public Accounts* (HMSO, London, 1973), p. x.

10. Ibid., p. x.

11. Kenneth W. Dam, *Oil Resources* (University of Chicago Press, London, 1976), p. 4.

12. Turner, 'State and Commercial Interests', p. 95.

13. *First Report*, p. xii.

14. D. I. Mackay and G. A. Mackay, *The Political Economy of North Sea Oil* (Martin Robertson, London, 1975), p. 25.

15. *Operations*, p. 49.

16. *First Report*, p. xv.

17. Ibid., p. xvii.

18. Ibid., p. xvii.

19. Edward N. Krapels, *Controlling Oil: British Oil Policy and the British National Oil Corporation* (US Government Printing Office, Washington, DC, 1977), p. 14.

20. *First Report*, p. xxi.

21. Mackay and Mackay, *Political Economy of North Sea Oil*, p. 32.

22. Jon R. Morgan, 'The Promise and Problems of Petroleum Revenue Tax' in *The Taxation of North Sea Oil* (Institute for Fiscal Studies, London, 1976), pp. 1–28.

23. Mackay and Mackay, *Political Studies of North Sea Oil*, p. 39.

24. Ibid., p. 39.

25. *First Report*, p. xxxiii.

26. Turner, 'State and Commercial Interests', p. 107.

27. Ian Smart, 'The Political Implications of North Sea Oil and Gas' in Saeter and Smart, *The Political Implications of North Sea Oil and Gas*, pp. 145–57.

28. Dam, *Oil Resources*, p. 104.

29. Mackay and Mackay, *Political Economy of North Sea Oil*, p. 39.

30. *Ilandføring av petroleum fra Ekofisk-området*, St. meld. nr. 51 (1972–3), (Ministry of Industry, Oslo, 1973).

31. *Petroleum Industry in Norwegian Society*, Parliamentary Report No. 25 (1973–4), (Ministry of Finance, Oslo, 1974).

32. *Operations*.

33. *The Scandinavian Oil and Gas Magazine*, no. 1/2 (1974).

34. Letter from the Norwegian Shipowners' Association to the Finance Committee of the Norwegian Parliament, cited by Øystein Noreng, 'Oljen og Norges politiske økonomi' in *Kontrast*, no. 3/4 (1974), pp. 11–24.

35. Dam, *Oil Resources*, p. 69.

36. *United Kingdom Oil and Gas Policy* (Department of Energy, London, 1974).

37. Turner, '*State and Commercial Interests*', p. 101.

38. *The Economist*, 26 July – 1 Aug. 1975, ' "Make or Break?" A survey of the North Sea', p. 6.

39. Mackay and Mackay, *Political Economy of North Sea Oil*, p. 32.

40. Dam, *Oil Resources*, p. 123.

41. Ibid.

42. Turner, 'State and Commercial Interests', p. 100.

43. 'Konkurrerer norske selskaper hverandre ihjel?' in *Norsk Oljerevy*, no. 6 (1978), p. 8.

44. *The Economist*, 5 August 1978.

45. *The Challenge of North Sea Oil* (Her Majesty's Stationery Office, London, 1978), pp. 3 ff.

46. *Petroleumsundersøkelser nord for 62° N*, St. meld. nr. 91 (1975–6), (Ministry of Industry, Oslo, 1976), p. 17.

47. *World Energy Outlook* (OECD, Paris, 1977), p. 47.

48. *Bravoutblåsingen – Aksjonsledelsens rapport* (Oslo University Press, NOU, Oslo, 1977), pp. 57 ff.

49. 'Norwegian Nightmare' in *The Economist*, 1 April, 1978, pp. 81–91.

50. Ibid., p. 81.

51. *Tillegg til langtidsprogrammet 1978–1981*, St. meld. nr. 76 (1977–8), (Ministry of Finance, Oslo, 1978), pp. 12 ff.

52. Amund Utne, 'Departement-offensiv for bedre eksport-prognoser' in *Sosialøkonomen*, no. 5 (1978), pp. 28–30.

53. ' "Statfjord B" med ny kostnadsrekord' in *Norsk Oljerevy*, nr. 6 (1978), pp. 14–17.

3 THE ECONOMIC ATTRACTION

North Sea Geology

The North Sea and the neighbouring land areas of Denmark, Germany, the Netherlands and the UK make up one geological unit, a sedimentary basin with clearly defined structural borders.[1] To the north-west of the basin is the Caledonian folding, which makes the mountains of Ireland, Scotland and Norway. The north-eastern limit is the Baltic-Scandinavian shield, and the southern edge of the basin is the Hercynian range. To the north the region is bounded by the deep-sea basin of the Atlantic ocean.

The geological age of these outer limits varies. The Baltic-Scandinavian shield is assumed to be more than 600 million years old, the Caledonian folding is believed to have an age of 400 million years, and the Hercynian range is thought to be 'only' 250 million years old. Thus the North Sea basin is geologically rather old. This contrasts with the adjoining deep-sea Atlantic basin, which divides the continental shelves of Norway and Greenland and is thought to be much younger. It is estimated to be 60 million years old and is probably the result of the separation of the Eurasian and American continents.

The North Sea basin has not been subject to any regional foldings. However, there have been several axes. Most of these structures were created from 350 million to 65 million years ago. The most important of these structures are the Norwegian Platform, stretching from north to south, and the Mid North Sea High or Ringkøbing-Fyn-High, stretching from east to west roughly along 56° N. The east-west structure divides the North Sea into two local basins. The southern basin is further subdivided by the Netherlands Ridge into the Anglo-Dutch basin to the west and the German basin to the east. The northern basin, which is generally deeper, is more complex, consisting of two platforms, the Norwegian one and the Orkney-Shetland one, and several grabens or troughs, the most important of which are the Mid North Sea or Central Graben, the Forties Trough and the West Norwegian Trough. These geological differences are of considerable economic importance, as the different structures have had different possibilities of accumulating petroleum through the geological age considered.

The common assumption is that petroleum is formed by the decay of marine plants and animals in shallow water, and that, in order to accumulate, the layers of oil must be trapped in porous rock that is covered over by non-porous rock, to keep the oil and gas from leaking out. Therefore, for oil to be formed in an area it must have been at one time covered by water and abandoned plant life, followed by porous rock and finally by a cap-rock.

In the North Sea there is a large area with many suitable traps for petroleum known as anticlines. There are systematic differences in the contents, probably because of differences in water coverage and marine life. The early part of the North Sea was probably land from which large rivers were flowing into deltas in the southern part. The climate was hot and humid, favouring abundant organic life in the fresh and salt water of the southern part. This was in the Carboniferous period, 600 million to 300 million years ago. Later, in the Permian period, 280 million to 220 million years ago, the rise of the Hercynian range might have provoked a general depression of the North Sea basin, with the ocean flooding in from the north on several occasions, shaping layers of sand, salt and chalk. Climate in this period was hot and dry with little organic life. Still later, in the Trias and Jura periods, 225 million to 135 million years ago, there were several local faultings. The southern part was definitely covered by salt water, furnishing rock-salt as a cap-rock.

In the Cretaceous period, 135 million to 65 million years ago, the climate was hot and humid with a high level of organic activity. The northern part was now covered by the ocean. Chalk sedimentation took place in the southern part and in the northern part the sedimentation in clay and sand were more dominant. Later still, in the Tertiary period, from 65 million to 1 million years ago, the climate gradually became colder. Chalk sedimentation continued in the southern part, while in the northern part troughs were filled with thick layers of clays, sands and shales.

This brief review of geological history explains why gas is mainly found at great depths in the southern part of the North Sea, while in the northern part oil and associated gas are found at more shallow depths. It is often assumed that gas alone occurs where marine life has been decaying in fresh or brackish water, and that oil and associated gas occur where marine life has been decaying in salt water. Also, the gas in the southern part of the North Sea seems to be fairly old in geological terms, while the oil and associated gas from the northern part seem to be of rather recent geological origin.

Most of the oil and associated gas have so far been found in the

central trough stretching from north to south between Scotland and
Norway, where sediments are probably thickest. So far, because of the
small amount of exploratory drilling, little is known about the oil
potential of the Shetland-Orkneys Platform and the West Norwegian
Trough, which is adjacent to the Norwegian Trench. The Norwegian
continental shelf north of 62° N has been subject to fairly detailed
seismic and geophysical surveys. Along the Norwegian coast there seems
to be a sediment of the Tertiary age that increases in thickness with
distance from the coast. The continental shelf off northern Norway
appears to be part of a large sedimentary basin in the Barents Sea that
could be very interesting from a petroleum point of view.[2]

Petroleum Potential

Since the discovery of gas in Groningen in the Netherlands the estimates
of the petroleum potential of the North Sea have varied tremendously.
The northern part was considered less promising until oil was found,
and since then it has been thought of as rather promising. It should also
be borne in mind that a resource only exists in relation to a price, and
the term 'reserves' is generally used to define resources available at
present prices and technologies. In this perspective the potential of
petroleum reserves has received a considerable boost by the price rise
in the early 1970s, and it can be further helped by technological
breakthroughs that may take place in offshore exploration and
production in years to come.

Because exploratory drilling is much more difficult and more
expensive at sea than on land, it is hard to assess the petroleum
potential of the North Sea exactly. On the one hand it can be argued
that given difficulties of exploration and production, and given the fact
that the rate of recovery is normally lower for an oilfield at sea than on
land, it is prudent to make conservative estimates of the petroleum in
given geological structures. On the other hand, it can also be argued
that oilfields are independent occurrences and that the degree of
uncertainty about petroleum potential decreases with the number of
fields.[3] The problem is that relatively small differences in judgement
can cause rather large discrepancies in estimates of potential reserves,
and this in turn influences company and government policy.

A few years ago this problem created a dispute. The British-Dutch
expert, Peter Odell, asserted that Western Europe could be self-
sufficient in oil, or even a net exporter, at a price much lower than

Figure 3.1: Basic Geological Structures of the North Sea Area

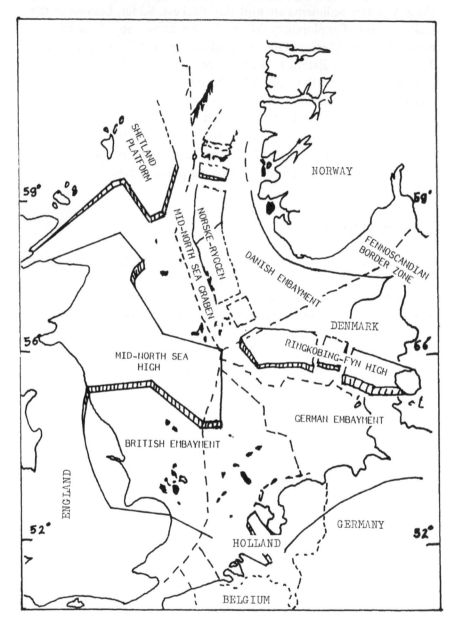

Source: *Operations on the Norwegian Continental Shelf,* Report No. 30 to the
Norwegian Storting (1973–4) (Ministry of Industry, Oslo, 1974).

Figure 3.2: Geological Times and Discoveries of Oil and Gas

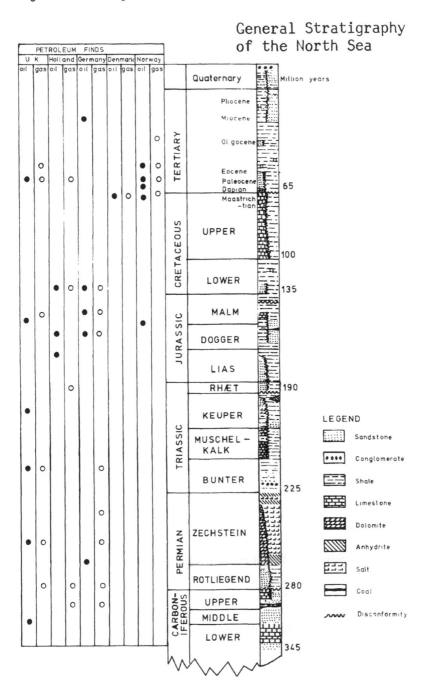

General Stratigraphy of the North Sea

OPEC's by accelerating the development of North Sea oil.[4] The method chosen to assess the potential reserves of the North Sea was to take published data on discovered fields and upgrade them by an analogy to fields elsewhere. The conclusion was that the ultimately recoverable reserves of North Sea oil alone, not counting the associated gas, were in the range of 10–11 billion metric tonnes (70–80 billion barrels).

Odell's method has met considerable criticism. First, there is the problem of comparing the recoverable potential of offshore fields with that of fields onshore. Techniques are different, which means different yields. Second, the North Sea in many ways seems to be a geologically unique area. More than 50 per cent of reserves are in a small number of large fields that were generally under very high pressure and present special problems of recovery. Third, with existing taxation, several fields that were either small or relatively deep tend to become economically marginal.

In response to this criticism one can reply that oilfields offshore are essentially similar to fields onshore, and that with improved techniques in the form of submersible completion systems or production from floating barges, which permit a more decentralised pattern of production, the rate of recovery could be improved considerably. Also, most of the oil in the world is found in a limited number of large fields.[5] This last observation also increases uncertainty, as other large fields could be found in areas that are still unexplored. It should be added that, from a geological point of view, there are indications of possible 'petroleum traps' in parts of the North Sea that have not yet been discovered.[6] In recent years reserve estimates for the North Sea have been revised upwards, and this trend could easily continue with more exploration, particularly in Norwegian waters. For example, if the geological structure close to Statfjord mentioned previously does contain the quantities of oil assumed by many, Norwegian oil reserves could easily double.[7]

Estimates for both the UK and Norwegian sectors can be considered as relatively conservative. This is particularly true of the Norwegian sector, where exploratory drilling so far has been rather limited. Much drilling remains to be done before both sectors are fully explored, and consequently reserve estimates could be increased substantially. On the other hand, it should be underlined that some of the most promising areas have already been explored, or are being explored. This is particularly true of the UK sector.

On the basis of a simulation model of geological frequencies, total recoverable oil reserves from the North Sea have been estimated to be

Table 3.1: UK North Sea Oil and Gas Reserves (licensed areas only)

	Oil (m tonnes)	Gas (billion cubic metres)	Total Oil (equivalents m tonnes)
Southern Basin			
— proven	—	513	513
— probable	—	113	113
— possible	—	65	65
— total	—	691	691
Northern Basin			
— proven	1,380	297	1,677
— probable	920	158	1,078
— possible	900	297	1,197
— total	3,200	752	3,952
Total British Sector			
— proven	1,380	810	2,190
— probable	920	271	1,191
— possible	900	362	1,262
— total	3,200	1,443	4,643

Note: Proven reserves are those which on available evidence are virtually certain to be producible at present prices and technology. Probable reserves are those which are estimated to have a better than 50 per cent chance of being producible at present prices and technology. Possible reserves are those which are estimated to have a significant but less than 50 per cent chance of being producible at present prices and technology. The amount of oil and gas in the category of possible reserves is by definition uncertain.

Source: *Development of the Oil and Gas Resources of the United Kingdom 1977* (Department of Energy, London, 1977), pp. 4—6.

Table 3.2: Norwegian North Sea Oil and Gas Reserves (licensed areas only)

	Oil (m tonnes)	Gas (billion cubic metres)	Total Oil (equivalents m tonnes)
Proven recoverable reserves	770	750	1,540

Source: *Fact Sheet: The Norwegian Continental Shelf* (Ministry of Industry, Oslo, 1977), p. 3.

in the range of 10,000–18,000 million tonnes (79–138 million barrels), but this estimate is necessarily quite uncertain.[8] If reserves are to be significantly expanded, this expansion will have to take place in the Norwegian sector.[9] Such an expansion should not be altogether excluded, as the Norwegian Petroleum Directorate estimates that total oil reserves south of 62° N are in the range of 3,000–4,000 million tonnes (22–29 million barrels).[10] In addition, there could be greater quantities of natural gas than are presently known to exist.

A moderately optimistic estimate for the Norwegian part of the North Sea south of 62° N would be oil reserves of 3,500 million tonnes and gas reserves of 2,000 billion cubic metres, making a total of 5,500 million tonnes of oil equivalents.

If UK reserves are estimated to be about the same or perhaps a bit larger, for example 4,500 million tonnes of oil and 2,500 billion cubic metres of gas, the two sectors combined would then have a potential of 8,000 million tonnes of oil and 4,500 billion cubic metres of gas, making a total of 12,500 million tonnes of oil equivalents. These figures seem reasonable and they are not too far away from the figure of 10,000–11,000 million tonnes of oil suggested by Professor Odell. However, it should be borne in mind that petroleum geologists have frequently been deceived in the past.

If the figures mentioned above do correspond to reality, the North Sea reserves will have a considerable impact upon the energy market of North-West Europe. The figure of 8,000 million tonnes of oil is not far from the proven and published reserves of Kuwait.

UK gas production is likely to increase from about 39,000 million cubic metres in 1977 to approximately 62,000 million cubic metres by the early 1980s. This level of gas production could be maintained for several years. UK oil production is likely to be around 100–120 million tonnes a year in the early 1980s (see Table 3.3).

Present discoveries have established the North Sea as an important petroleum province. Consequently, there are good statistical chances for assuming that continued exploration will give more discoveries and an expansion of reserves. However, the usual trend, observed in other oil provinces, is that the largest fields are found first, and that over time exploration will result in steadily smaller fields being found. With the relatively limited exploration carried out so far, especially in the Norwegian sector, chances are that large fields could still be found. Odds are that at least one more giant oil and gas field will be found in the North Sea.[11] There is a growing consensus that the ultimate reserves of oil in the UK sector of the North Sea may be about 3,000–3,500

Table 3.3: Forecast of UK North Sea Oil Production (m tonnes)

1978	60–70
1979	80–95
1980	90–110
1981	100–120

Source: *Development of the Oil and Gas Resources*, p. 3.

Table 3.4: Forecast of Norwegian North Sea Oil and Gas Production

	Million Tonnes of Oil Equivalents
1978	40
1979	60
1980	67
1981	65
1982	66
1983	65
1984	69
1985	72
1986	75
1987	70
1988	66
1989	60
1990	54

Note: These figures only concern production from the fields Ekofisk, Frigg and Statfjord. Other fields are likely to start producing in the early or mid-1980s.

Source: *Norwegian Long Term Programme 1977–1981* (Ministry of Finance, Oslo, 1977), p. 123.

million tonnes of oil, of which about two-thirds have already been found.[12] The remainder will essentially be in smaller fields, implying a substantial effort of exploration to be found, as well as high costs of development. The larger fields may essentially be found in the Norwegian sector, possibly bringing the Norwegian reserves up to the level of the UK ones. Thus, the North Sea could realistically contain 6,000–7,000 million tonnes of oil in addition to natural gas reserves corresponding perhaps to half of this. This makes the North Sea an oil province of less importance than, for example, Kuwait, and probably of less importance than Mexico.

The oil and gas reserves of the UK and Norway might be substantially expanded if large quantities are found in the new areas. On the UK

side, the new areas are to the west of Shetland and to the west of
Scotland as well as off Cornwall. On the Norwegian side, the new areas
are north of 62° N. In some of these areas, geological prospects for
finding oil and gas are considered good. Given the large size of the
areas, there is indeed a fair chance of finding petroleum. However,
exploration may be difficult and development more expensive than in
the North Sea, due to worse weather conditions or large depths. Also,
lead times could be long, and output could be held down because of
production limits. The conclusion is that the petroleum of the North
Sea has an immediate effect of making North-West Europe somewhat less
dependent upon OPEC oil, but it does not change substantially the energy
situation of Western Europe, not to mention that of the OECD area.

North Sea Oil in the World Market

The economic attractiveness of a given oilfield is determined by four
basic factors:

    the location in relation to markets;
    the quality of the crude oil;
    the costs of development and production;
    the security of supplies.

The oilfields of the North Sea satisfy three of these criteria very well.
They are located in the immediate vicinity of one of the world's major
oil markets, North-West Europe, and they are fairly close to the East
Coast of the United States. The crude oil found in the North Sea has
been of generally good quality, light with a low sulphur content, so it is
particularly well suited for the more expensive refined products, such
as naphtha and gasoline. These two factors increase the value of the
North Sea oil. The returns that result from these advantages can be
termed a locational rent and a quality rent.[13]
  The oilfields of the North Sea are also located within a stable
political context, belonging to industrial democracies that have close
ties with other oil consuming countries. Consequently North Sea oil
can be considered secure. If all other factors are equal, many consumers
would prefer to be dependent on oil from the North Sea rather than
comparable crudes from areas with less political stability, such as the
Middle East. The returns that result from the security of supply can
be considered a political rent.

In contrast to the locational rent and the quality rent which are relatively easy to identify and to quantify, the political rent can take several forms, which are not necessarily readily quoted in the market. However, it can be expressed commercially — for example, a secure supplier can demand stricter conditions of payment. In a situation of slack demand, a secure supplier is more likely to keep customers. The political rent can also be expressed politically in the relationship between governments and companies. Oil companies can consider a stable political context an advantage from the point of view of risk minimisation, and therefore let political stability offset higher costs and possibly higher taxes. Also, governments that are stable suppliers can perhaps obtain benefits such as the transfer of technology or joint ventures more easily.

These three advantages of North Sea oil — favourable location, good quality and security of supply — perhaps explain the interest of the international oil industry in the area, despite the obvious drawback of very high development and production costs. In the late 1950s and throughout the 1960s the real price of oil was a good deal lower than production costs in the North Sea. Nevertheless, the international oil industry was starting to invest in oil production that probably would cost between $2 and $6 a barrel at a time when the price of oil was $2 a barrel or less. The reason for this behaviour was probably growing pessimism in the oil industry about expanding reserves in traditional, cheaper areas of production.[14] Long before the oil crisis there was evidently a fear that production in the Middle East could not be expanded beyond a certain ceiling, and that this ceiling would be reached early enough to make investment in the North Sea worth while.

Problems of Development

In the same way that Alaska is a pioneer province for petroleum development in Arctic conditions, the North Sea is a pioneer province for petroleum development on the offshore continental shelf outside tropical or semi-tropical areas. The only other similar offshore area of significance is the continental shelf off Eastern Canada, but the activity there has been on a much smaller scale.[15]

Offshore development has certain problems of its own, because of water and weather, which make development much longer and much more expensive.[16] The North Sea is rather shallow and water depths are in general not obstacles to development. On the other hand, weather conditions are such that drilling is normally feasible for only approximately 175 days a year.[17] These rough weather conditions require equipment that is exceptionally robust, and the installations to transport oil and gas must be well protected.

In these circumstances any exploration and production is a very risky undertaking. Workers can easily be injured or killed. The environment can be harmed by pollution. Equipment can be idle because of delays and can also be damaged or lost. One of the reasons why the North Sea has attracted much attention and capital in spite of these difficulties is that it is a test area. If successful, methods can be developed for use elsewhere, which cuts future lead times and costs. In a test area lead times and costs are generally high and escalating until a critical point is reached when the new technology is mastered. Then lead times and costs start decreasing again.[18] This is a general pattern of innovation which applies to the North Sea. An important result is that the production of oil and gas in the North Sea in the future can be expected to be less expensive and more profitable, even at constant oil prices. This explains why some companies are deferring development, which does seem to be the case in the Danish sector. From this perspective, the Norwegian decision to have a moderate rate of development, at least in the inital period, appears to be a wise one, unless of course the moderate rate of development postpones the critical point when the new technology is mastered. The need of the UK to develop its oil potential quickly might eventually be regretted because of the higher costs involved in early exploitation. If they can afford it, similar offshore areas around the world should wait until the British and Norwegians have mastered the technology and reduced costs and lead times.

The first step in development consists of seismic and geochemical surveys. Seismic surveys are relatively easy and inexpensive to undertake at sea. However, these surveys can only establish that certain geological conditions are present. The next step is to drill a well. At sea this can be done in four ways:

(1) drilling vessels, which can be used mainly in good weather conditions;
(2) submersible rigs, which can only be used in certain depths and weather conditions;
(3) jack-up rigs standing on legs on the bottom of the sea, which can only be used in shallow and sheltered waters,

(4) semi-submersible rigs, which float on the surface with a large substructure reducing the impact of waves.

This last type of rig is best suited for use in rough weather conditions, but they are expensive to build and operate, and must be serviced from bases on shore by helicopters and supply ships.[19] Semi-submersible rigs are capable of drilling in water depths of 300 metres (1,000 ft) down to at least 10,000 metres (30,000 ft).

When a field is deemed commercially exploitable, operators have few alternatives concerning development. The choice is essentially between a centralised or a more decentralised pattern of production. So far, the most common method has been to use large concrete platforms, a very centralised pattern of production. Such massive platforms have been thought necessary to resist the waves and winds of the North Sea. Norwegian industry developed the integrated type of concrete platform, Condeep, particularly for North Sea conditions. Large platforms are being used to develop the Forties field and the Statfjord field. However, such massive platforms are extremely expensive to build and a large part of the escalation of costs and lead times is connected with the platforms of this type. There is now growing interest in smaller production platforms, which are thought to be less complicated and less expensive. These would create a more decentralised pattern of production, spreading risks and vulnerability.

At present, development costs are estimated to make up close to 90 per cent of the total expenditure connected with a North Sea oilfield. Breaking down cost further, platforms generally make up between 50 and 70 per cent of total development expenses. The other dominant ingredient is the transportation system, in the form of pipelines or marine terminals. Current production costs are influenced by the fact that a large platform at sea requires a lot of personnel, and must be serviced from shore by supply ships and helicopters.

Finally, both development costs and production costs are influenced by safety requirements, environmental protection and labour regulations. For example, the decision of the Norwegian Petroleum Directorate that the B platform of the Statfjord field should be accompanied by a separate housing platform would, if maintained, have implied sizeable additional costs. Also, the new Norwegian regulations on working hours have been one of the factors behind the cost escalation on Statfjord.

Given the complexity, cost and lead times of large platforms, there is now considerable interest in new and relatively radical solutions to the problem of getting at the oil. One possible solution is subsea completion systems, with complete installations on the bottom of the sea connected by pipelines. Another solution envisions production installations on large barges connected to wells on the bottom by flexible pipes. In case of adverse weather conditions, the barges could be disconnected from the bottom installations. The oil would be brought from the barges to shore by tankers. Such barges could also be used for gas if they contained installations for liquefaction.

Such systems would have two major advantages. They could reduce production costs significantly and they could provide a better technical and economic basis for the exploitation of smaller fields. The British and Norwegians have had a preference for the large production platforms, and most of the research and development concerning new solutions appears to be going on in the United States.[20] Breakthroughs in these matters could cut the production costs of North Sea oil and would also permit a significant upgrading of the resource base.

Transportation essentially involves a choice between pipelines and tankers for oil. Oil transportation by pipeline generally is the best solution. High capital costs are easily offset by low operating costs, once a critical minimum level of throughput is reached. Transportation by tanker is less favourable because it is less regular in adverse weather conditions, and operating costs are higher, in part because of the small distances travelled to the markets of North-West Europe.

There are, however, several factors that complicate the pipeline solution. First, because of the importance of the North Sea to fishing, all North Sea governments require that pipelines be covered. This means that pipe cannot simply be laid on the bottom of the sea. A trench must be dug and covered again after the pipe has been laid down. Exposure of some miles of gas pipeline in the summer of 1977 caused the Danish government to forbid the use of the line from Ekofisk to Emden. This delayed deliveries of gas to West Germany and revenues to Norway. The Norwegian Trench offers a even more serious obstacle to pipelines. It is doubtful that existing technology can connect the Norwegian oilfields with Norway's coast by pipeline, especially if economic return and safety considerations are taken into account.

Safety and environmental considerations must be taken much more seriously with offshore oil production. Water and weather make installations much more vulnerable to accidents. First of all, there is the danger to workers mentioned above. A large production platform can have from 50 to 200 people working at any one time. If a blow-out is followed by an explosion, lives are bound to be lost. This was the background of the decision by the Norwegian Petroleum Directorate to demand a separate housing platform at Statfjord B.

The North Sea is one of the world's roughest bodies of water, and rigs must be constructed to withstand waves that statistically are thought to occur only once every 100 years. These can be waves 30 metres (100 ft) in height. In one case a rig experienced a 'hundred-year wave' only a few weeks after entering the North Sea.[21] Deficiencies of construction can also cause accidents. One rig was wrecked because of brittle fractures in hangers holding the deck.[22] Human errors in organisation and inspection are also a potential source of accidents. The Bravo accident, until 1980 the most serious one in the North Sea, was essentially a result of human neglect.[23]

The danger of oil pollution to marine life is a matter of intense dispute, and relatively little is known about it. Considerable quantities of oil leak into the sea from platforms, loading installations, tankers and pipelines. In addition, a blow-out could cause thousands or even millions of tonnes to flow into the sea. For example, if a platform catches fire during a blow-out, a new well must be drilled to relieve the pressure on the platform where the blow-out is. This process could take weeks or even months, particularly if weather conditions are adverse. During this time perhaps millions of tonnes of oil could flow into the sea. Some maintain that this could cause serious environmental damage. The oil is likely to create a thin film on the top of the sea, which would be very harmful to sea birds. It could also harm marine life, as the upper layer of the sea is most important to its reproduction.[24]

On the other hand, apart from damage to sea birds, very few negative effects of oil pollution have so far been proved in practice. As mentioned earlier, there were no significant negative effects on marine life from the oil spills of the Bravo blow-out.[25] It should also be added that these spills were of very small proportions. Laboratory tests have shown that plankton, shellfish, fish eggs and larvae are very sensitive to oil.[26] In general, the impact of oil could be particularly harmful on spawning grounds.[27]

Oil activities offshore are accompanied by safety and environmental hazards that can be reduced, but never entirely eliminated. The regulations reducing the hazards generally imply higher costs of development and production for the oil industry. Consequently, any decision to develop a North Sea oilfield implies a trade-off between oil and fishing interests. The stronger the fishing interests, the stricter the regulations and the higher the costs for the oil industry. This is particularly relevant for future oil development off northern Norway, where several of the best fishing banks are also promising from a petroleum point of view.

Cost Factors

Production costs vary considerably among the fields of the North Sea. Little detailed information on costs for the different fields is available, but some generalisations are still possible.

Among the factors related to individual fields, size is the most important.[28] Development costs normally do not increase as a function of field size, which means that the ratio between capital and output decreases with the size of a field. In addition, there are obvious economies of scale.[29] For example, several wells can be drilled from the same platform, and servicing the offshore operations can be simplified with increasing size. Finally, at a certain size transportation by pipeline becomes far more economical than transportation by tanker.

The productivity of the reservoir is another important variable. This of course depends on the pressure of the field, the permeability of the rock that contains the oil, and the viscosity of the oil itself. Pressure decides the time when gas or water must be injected into the field in order to keep oil flowing. As with the Ekofisk field, some fields can have increasing production of gas even after the production of oil starts declining.[30]

Water depth is another critical factor affecting production costs.[31] The technical complexity and the costs of exploratory drilling and development increase with water depth. Traditional technology can drill in only very shallow waters. Some wild-cat drilling has been done at depths of 500 metres (1,500 ft), but present technology seems to have a limit of 200–250 metres (600–750 ft) for commercial oil production.[32] Some of the North Sea fields are close to the limits of present technology, and production costs escalate very rapidly in depths between 100 and 200 metres. The Statfjord field, so far the largest North Sea field discovered, stands in about 150 metres (450 ft) of water,[33] making it very expensive to exploit.

The distance from shore is of great importance for two reasons. First, the cost of servicing rigs by helicopters and supply ships is directly related to the distance travelled from the base on shore. Second, transportation costs are also influenced by distance.

Local weather conditions also have important effects on relative costs between various North Sea fields. The northern part of the North Sea has much rougher weather conditions than the southern part. Weather conditions on the western coast of Scotland and off northern Norway are generally considered to be rougher than those in the North Sea. In the rougher areas drilling is more limited, equipment must be made more resistant, and servicing by supply ships and helicopters is harder.

As already mentioned, it is generally far less expensive to transport a given quantity of oil by pipeline than by tanker, but a certain critical level of yearly throughput must be achieved in order to make the pipeline economical. Pipelines also have exceptional economies of scale because, with enlargement of the diameter, capacity expands more quickly than do capital and operating costs.[34] The North Sea is a frontier territory with regard to oil pipelines. Weather conditions complicate the laying of pipelines and increase capital costs, and costs are further increased by the requirement that pipelines be buried. The total cost of the system of pipelines from Ekofisk to Tees-side (oil) and Emden (gas), about 750 kilometres of large-diameter pipeline, is estimated to be at least $700 million, or $1 million per kilometre (in 1974 dollars).[35] However, the low operating costs make the pipeline alternative attractive in spite of these high capital costs. Pipelines have been laid from the Frigg gas field and from several UK oilfields to the UK coast. The existence of a nearby pipeline can make many marginal fields commercially exploitable. This is one of the reasons why both the UK and Norway are at present considering the construction of large pipeline systems, particularly for gas. These pipelines would tap many smaller fields in addition to transporting oil or gas from some of the larger fields.

To sum up, all North Sea fields are distinct, with their own special characteristics and economic advantages and drawbacks. Therefore it is practically impossible to assume a general cost level for North Sea oil and gas production. Ultimately, every field must be judged independently.

Potential Return

Because of the variations of costs between different fields (see Tables 3.5 and 3.6), any generalisation about the economic benefit of North Sea oil is likely to produce a distorted picture. For example, the average investment cost has escalated and is now perhaps around $1.50/barrel

Figure 3.3: Possible Breakdown of Expenditure for Developing a North Sea Oil and Gas Field

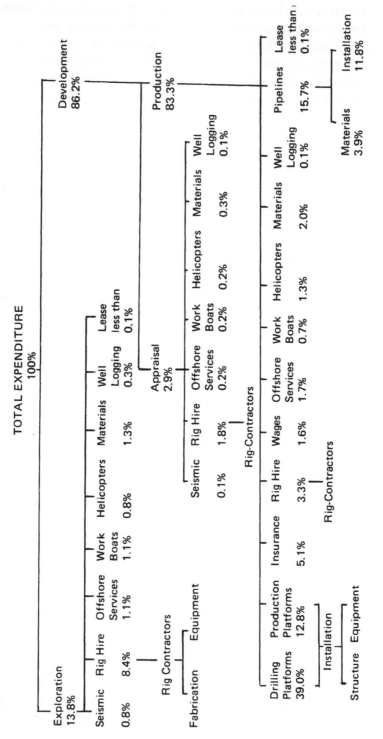

Source: Cazenove and Co., *The North Sea, The Search for Oil and Gas and the Implications for Investment* (Cazenove, London, 1972).

Table 3.5: Estimated Development Costs of Selected North Sea
Oilfields (£m)

	British Field			
	Argyll	Auk	Piper	Forties
Production platforms	—	20 (50%)	65 (3½%)	380 (58%)
Rig conversion	2 (18%)	—	—	—
Single-buoy mooring	2 (18%)	3 (8%)	—	—
Pipelines	—	—	65 (30%)	110 (17%)
Terminals	—	5 (13%)	50 (23%)	40 (11%)
Development drilling	4 (36%)	8 (20%)	20 (9%)	65 (10%)
Other expenditure	3 (27%)	4 (10%)	20 (9%)	55 ((8%)
Total	11	40	220	650

Source: D. I. Mackay and G. A. Mackay, *The Political Economy of North Sea
Oil* (Martin Robertson, London, 1975), p. 71.

Table 3.6: Estimated Development Costs of the Norwegian Ekofisk
Field

Item	Cost (m Nkr.)	Cost ($m)
Platform, equipment, development drilling Ekofisk field (main field)	1,485	270
Total development Tor field (related), incl. platform, dev. drilling, local pipeline	385	70
Total development West Ekofisk field	220	40
Total development Cod field (related)	440	80
Ekofisk centre, tank, processing equipment, platform, tank deck, etc.	1,512	275
Pipeline to Tees-side (oil)	1,650	300
Pipeline to Emden (gas)	2,200	400
Shore installations Tees-side	1,485	270
Shore installations Emden	330	60
Total	9,707	1,765

Source: *Operations,* p. 14.

(1977 prices). But some fields are far more expensive; for Statfjord the investment cost is between $2.00 and $2.50/barrel. Other fields, particularly those developed early in the 1970s, such as Argyll, Auk, Piper, Forties and Ekofisk, have much lower investment costs, generally less than $1.00/barrel. Average operating costs were about $1.20/barrel in 1977–8; so production costs in the North Sea average $2.50–$3.00/ barrel, with large variations between fields.[36] By comparison, production costs in the Middle East are thought to average $0.10/0.15/ barrel.[37]

Calculating capital and operating costs presents some problems of method. A rather simplistic way is to take the total capital or operating costs over the life of the field divided by the volume of the output.[38] This procedure ignores the timing of production and revenues in relation to capital outlays, and the interest on capital invested.

Table 3.7: Estimated Capital and Operating Costs for Four Selected Fields.

Field	Capital Costs ($m)	Operating Costs ($m)	Output (thousand barrels)	Capital Cost/ Barrel ($)	Operating Cost/ Barrel ($)	Total Cost/ Barrel ($)
Argyll	25	330	156,950	0.16	2.10	2.26
Auk	90	190	102,220	0.88	1.55	2.43
Piper	500	460	790,225	0.58	0.54	1.12
Forties	1,500	1,455	1,579,355	0.95	0.93	1.88

Source: Mackay and Mackay, *The Political Economy of North Sea Oil*, pp. 46–7 and 97.

Table 3.7 indicates that the relatively small fields of Argyll and Auk have correspondingly low capital costs and high operating costs. The Piper and Forties fields, which are relatively large, have correspondingly high capital costs and low operating costs. Piper and Forties are more typical of the oilfields of the North Sea, where operating costs account for less than half or perhaps nearly a third only of total production costs.

The recent escalation of capital costs and lead times meant that total production costs per barrel of oil have increased. This also implies that the total economics of North Sea oilfields are increasingly sensitive to capital costs. Consequently, technological breakthroughs could potentially have a very significant impact.

The revenues and economics of a field also depend upon the

production profile chosen. As a general rule, the longer the time horizon of an oilfield, the higher the rate of recovery and the total output. Consequently, from the cost point of view it can be preferable to have a stretched-out production profile. This in turn must be offset against capital costs and the cash flow desired. When the production profile of the Norwegian Ekofisk field was decided on, two alternatives were considered, one having a high peak relatively early and the other having a stretched-out profile with more even volumes for a longer number of years. The latter was chosen in part because it was thought that the first profile would create difficulties for marketing the gas.[39]

When considering revenues, oil and gas should be strictly separated. Oil prices in North-West Europe are essentially determined by the world oil market. Traditionally the Rotterdam spot market has handled 2–3 per cent of the world's oil transactions. In recent years, because of the nationalisation of oil operators in many countries, a relatively greater proportion of the world's oil trade tends to be subject to open market transactions, and consequently the overall importance of the spot markets such as Rotterdam is increasing. As previously mentioned, the North Sea crudes have additional market value because of their favourable location and good quality. In practice, the selling price tends to be established by the governments concerned within a range determined by the market for comparable crudes.

Gas prices deserve special mention, as there is no international market for gas. Problems of transportation fragment the gas market into many local units with very different prices. Gas produced in the UK sector has to be landed in the UK, where the British Gas Corporation is the only buyer. Consequently, the price of the gas results from complete negotiations between the British Gas Corporation and the companies taking into account the gas from each field.[40] Generally, the price is not disclosed, but it is reported to have increased considerably from the early agreements in 1968 to later agreements in the mid-1970s.[41] Gas from the partly Norwegian Frigg field has probably received a higher price from the British Gas Corporation than has been paid for gas from UK fields.[42] The bargaining position of the British Gas Corporation was of course much weaker with companies operating in the Norwegian sector, because these firms were not compelled to land the gas in the UK and could threaten alternative solutions.[43]

Generally, gas landed in the UK from the Southern British sector in the North Sea had a price of approximately 1.2–1.5 pence per therm in the late 1960s and early 1970s, and up to 3.4 pence per therm after the oil crisis.[44] Gas landed from the Norwegian sector to the Continent of Europe generally has a price that is tied to the price of oil, with provisions for increases. The price of Ekofisk gas f.o.b. Emden was agreed upon to be Nkr. 0.14 per cubic metre or US$0.52 per 1,000 cubic feet with provisions that should be regularly adjusted according to the development of the price of certain types of oil on the Continent from 1972.[45]

Given this basic knowledge about production costs, it is possible to make a general estimate of the return on investment in North Sea oilfields at different oil prices. We get a rough idea of the returns by looking at the gross revenues generated by the field before the payment of taxes or royalties to the UK or Norwegian government. In the Argyll field, a relatively favourable venture, production started in the first year of development and capital costs were particularly low because a drilling rig could be converted into a producing unit.

Table 3.8: Simulated Cash Flows for the Argyll Field

Year	Capital Costs ($m)	Operating Costs ($m)	Output (thousand b/d)	Cash Flow at $3/b ($m)	Cash Flow at $12/b ($m)
1974	25	10	10	− 24	7
1975	−	40	30	− 7	92
1976	−	40	60	26	223
1977	−	40	60	26	223
1978	−	40	60	26	223
1979	−	40	60	26	203
1980	−	40	50	15	179
1981	−	40	50	15	179
1982	−	40	50	15	179

Note: Net present value (15 per cent) of cash flow at $3/b: $64 m. Net present value (15 per cent) of cash flow at $12/b: $843 m. Internal rate of return of cash flow at $3/b: 71.4 per cent.

Source: Mackay and Mackay, *The Political Economy of North Sea Oil*, p. 46, on the basis of data supplied by Wood, Mackenzie and Co., *Political Economy*.

The Piper and Forties fields are more typical of the rest of the North Sea fields. The Argyll field, small, easy to develop and offering an easy transportation solution, was a good deal even at pre-1974 prices. The development of the Piper field could be justified by its total economics, but the rate of return on capital invested, even before royalties and taxes, would have been less than what was usual in the oil industry. The economic sense of developing the Forties field at pre-

Table 3.9: Simulated Cash Flows for the Piper Field

Year	Capital Costs ($m)	Operating Costs ($m)	Output (thousand b/d)	Cash Flow at $3/b ($m)	Cash Flow at $12/b ($m)
1973	100	—	—	− 100	− 100
1974	200	—	—	− 200	− 200
1975	200	10	20	− 188	− 122
1976	—	30	100	80	408
1977	—	30	225	216	956
1978	—	30	225	216	956
1979	—	30	225	216	956
1980	—	30	225	216	956
1981	—	30	202	191	854
1982	—	30	182	169	766
1983	—	30	164	150	688
1984	—	30	148	132	618
1985	—	30	135	118	562
1986	—	30	121	103	500
1987	—	30	109	89	446
1988	—	30	98	77	399
1989	—	30	88	66	355
1990	—	30	80	58	320

Note: Net present value (15 per cent) at $3/b: $291 m. Net present value (15 per cent) at $12/b: $2,863 m. Internal rate of return at $3/b: 27 per cent. Internal rate of return at $12/b: 84 per cent.

1974 prices was unquestionable, assuming a stable price of oil. Still, the development of these two fields, and of others, including the Norwegian Ekofisk field, which in some ways resembles Forties, was commenced prior to 1974.

Consequently, the decision to develop the North Sea oilfields in the early 1970s must be seen in a dynamic perspective. Most likely, the investment decision was not based on calculations using a static price of oil, but rather on a well founded expectation that the price of oil in real terms would rise substantially over the lifetime of the fields concerned, in some cases 25 years, enough to secure a handsome profit. It is telling that exploration had been active in the ten years preceding the 1973–4 price rise.[46] Exploration is the real risk, and there is little reason to think that the oil companies would have been willing to risk capital for exploration purposes if the potential finds would be uneconomic to develop in a long-term perspective. This indicates that the oil companies active in the North Sea in the 1960s and early 1970s had foresight and a dynamic understanding of the situation, as well as

Table 3.10: Simulated Cash Flows for the Forties Field

Year	Capital Costs ($m)	Operating Costs ($m)	Output (thousand b/d)	Cash Flow at $3/b ($m)	Cash Flow at $12/b ($m)
1972	150	–	–	– 150	– 150
1973	300	–	–	– 300	– 300
1974	600	–	–	– 600	– 600
1975	300	30	50	– 275	– 111
1976	150	75	200	– 6	651
1977	–	75	400	363	1,677
1978	–	75	400	363	1,677
1979	–	75	400	363	1,677
1980	–	75	360	319	1,502
1981	–	75	324	280	1,344
1982	–	75	292	245	1,204
1983	–	75	262	212	1,073
1984	–	75	236	183	959
1985	–	75	213	178	937
1986	–	75	191	134	762
1987	–	75	172	113	678
1988	–	75	155	95	604
1989	–	75	139	77	534
1990	–	75	126	63	477
1991	–	75	113	49	420
1992	–	75	102	37	372
1993	–	75	92	26	328
1994	–	75	82	15	284

Note: Net present value (15 per cent) at $3/b: $159 m. Net present value (15 per cent) at $12/b: $3,783 m. Internal rate of return at $3/b: 12 per cent. Internal rate of return at $12/b: 54 per cent.

having a framework of reference extending over a long period of time. It is questionable to what extent the governments, on whose territories the activities took place, had a similar understanding of the situation.

At the post-1973 oil prices the economic attraction of the North Sea oilfields changed radically. Some fields, like Argyll, became exceedingly profitable. Other fields, like Piper, Forties and Ekofisk, whose development had commenced in the early 1970s, became highly profitable by the usual standards of the oil industry. Calculations indicate that at a price of $11.00/b and with the historical British regime of oil taxation, i.e. royalties at 12.5 per cent and a corporation tax of 52 per cent, the internal rate of return would have been 279 per cent for the Argyll field, 52 per cent for Piper and 33 per cent for

Forties.[47] In its proposal to introduce the special tax on petroleum extraction, the Norwegian Finance Ministry examined the economics of two hypothetical oilfields at the new price of oil.[48] In the first example, which seems to be a relatively small field, with a peak production of perhaps 2.6 million tonnes a year (52,000 b/d) during ten years, with capital costs of about $500 million and operating costs of generally about $22 million a year, the internal rate of return is supposed to be 27.5 per cent in the case of full self-financing, and in the case of 75 per cent external financing, at an interest of 12 per cent a year, the internal rate of return is supposed to be 37.3 per cent. In the second example, which seems to be a fairly large field, not too different perhaps from Ekofisk, with a peak production of about 20 million tonnes a year (400,000 b/d), with capital costs of about $2,500 million and operating costs of about $120 million a year, the internal rate of return is supposed to be 52.3 per cent in the case of full self-financing, and 75 per cent in the case of 75 per cent external financing. This return on investment made the North Sea appear as one of the most profitable areas available for the international oil industry, compared with historical records as well as with other areas. In the early 1950s the rate of return on investment by the international oil industry outside the United States had approached 30 per cent.[49] Since the late 1950s the rate of return had declined drastically, and on the eve of the price rise in 1973–4 it was hardly more than 10 per cent on average.[50] Also, in most other oil producing areas outside North America, the trend was increasingly in the direction of full nationalisation of oil production, leaving the international oil companies as traders, transporters and refiners of oil produced by the state oil companies of the producing countries.[51] This made the North Sea appear one of the relatively few areas where the international oil industry could get access to new equity oil with both a high degree of security of supply and potential for considerable profits.

The distinction made by the Norwegian Finance Ministry between fields developed through equity capital and through partial external financing has a practical relevance. Historically, the international oil industry has been largely self-financing.[52] Exploration and development have usually been financed with equity funds, i.e. by capital accumulated within the companies. In the late 1960s and in the 1970s falling profit margins and rising costs have made the international oil industry rely increasingly upon external sources of finance. By 1972–3, i.e. before the price rise, the international oil companies probably only covered about 70 per cent of their investment costs through equity financing.[53] This proportion seems to be falling considerably in the 1970s, as exploration and production move into more difficult areas, and as profit margins are reduced because of increasing nationalisation in the OPEC countries. In the North Sea, the rate of external financing could in some cases be up to 90 per cent.[54] In some fields, which are especially expensive or risky to develop, the trend is to engage a higher proportion of capital than is the case on average. The major risk is linked to exploration, so once a field has been found and declared commercial, the risk is essentially linked to costs escalating and delays in production. The external financing does not increase these risks, and as shown in the Norwegian examples, it can have a favourable impact on cash flows and rates of return, implying a transfer of income from the government-landowner to the concessionary companies, provided that interests on debts can be deducted from taxable income. Thus, increasing external financing can further improve the economic attraction of the North Sea oilfields. The international capital market has in general shown a considerable interest in financing North Sea oil development, in spite of rising costs.

This general impression of good economics should not hide the fact that there are substantial differences between fields, and consequently it is difficult to generalise costs and returns. This complicates the task of designing a taxation regime which can capture a given part of the economic rent for the government, without too unequally affecting the economics of the different fields. Another dimension of difficulty is linked to the development of the price of oil over time. It is difficult for governments explicitly to base taxation and fiscal regulations on anything but current prices, i.e. seeing the problem of capturing the rent in a static perspective. Oil companies, however, can in their own calculations base scenarios for revenue, cash flows and rates of return on the assumption that the price of oil is likely to rise, i.e. in a dynamic perspective. Thus, within the framework of the concessionary system, the premisses of economic calculation are unequal between governments and oil companies. It should also be mentioned that the cost and price movements have a political significance, in the sense that they affect the bargaining relationship between governments and companies. With improving economics, as in the aftermath of the price rise in 1973–4, the position of governments improves. With deteriorating economics, as a result of a price drop or a cost escalation, the position of governments deteriorates.

Cost Escalation and Industrial Organisation

The escalation of production costs and lead times in the North Sea has
been particularly strong since the oil crisis of 1973. This might lead one
to suspect that the escalation of costs, to a certain extent at least,
amounts to transfer of oil profits from operating companies and
governments to other parts of the industry. Without excluding such an
eventuality, it must be pointed out at the outset that the situation is
more complex.

An escalation of costs and lead times is a fairly common phenomenon in large industrial projects. The problem in the North Sea is that
the cost escalation in recent years has been unusually strong, and in
some cases quite dramatic. In the UK, simple indices of cost estimates
made at different intervals show costs rising at an annual rate of up to
90 per cent for a sample of projects.[55] In Statfjord, the principal
Norwegian field being developed after 1974, development costs have
risen two or three times in a couple of years.[56] In the late 1970s this is
the general picture in the North Sea. If extrapolated, this cost trend
could make North Sea oil unprofitable for private companies in the
1980s. However, for reasons that are explained below, such a trend
seems fairly unlikely.

The development of gas fields in the southern part of the North Sea
did not suffer from the same increases in costs and lead times. This
led to a feeling of confidence among planners in governments and oil
companies, and it was widely believed that the problems of development were largely mastered. As operations moved further north, costs
rose and lead times grew longer, but the lags of performance were not
large enough to receive widespread attention. This was the case with
the UK Piper field and the Norwegian Ekofisk field, both situated at
the south edge of the northern part of the North Sea. Only as operations
moved into more northern waters, after 1974, with fields such as Brent
on the UK side and Statfjord on the Norwegian, did performance start
to lag significantly behind anticipations.

Presently costs are rising and lead times are being stretched out, but
the cost escalation is in total not much worse, if at all, than that of
several other industrial projects onshore. Petrochemical plants in the
UK, Norway and other countries often exceed their estimated costs by
50 to 100 per cent. A major UK steelworks project in the 1970s
exceeded its anticipated cost by 78 per cent.[57] UK electricity plants
constructed in the 1970s have had costs slipping by more than 200
per cent in excess of original estimates. Finally, the estimated cost of

the Concorde project increased from £170m in 1962 to £1,065m in 1973. This record is still unbeaten by any North Sea oil project. The point is that industrial projects, with a high content of innovation and development, tend to be characterised by a much higher cost escalation than projects based on a prototype which again is the result of broad experience.[58] Because the North Sea oil projects have been pushed through in order to achieve a quick completion of development, an early start of production and a high return on capital, the cost escalation has been compressed in time and may appear high measured on a year by year basis. Measured on a comparable project basis, the cost escalation is of more normal proportions.[59]

Several reasons can be pointed out why development problems, resulting particularly in rising costs but also in longer lead times, have become more acute since 1974.

(1)   Only after 1974 did operations start in the northern waters, where depths are greater and weather conditions are worse than further south; a critical limit for operating conditions, due to a deterioration of both depths and weather conditions, seems to have been reached after 1974, between Ekofisk and Statfjord on the Norwegian side, and between Piper and Brent on the UK side.

(2)   The management of the projects, based largely on the experience of the international oil companies around the world, which had proved fairly satisfactory in the more southern waters of the North Sea, has proved quite unable to cope in a satisfactory way with large projects with high contents of innovation and development.

(3)   When encountering new problems in projects with high contents of innovation and development, there is a trade-off between a time-conscious strategy and a cost-conscious one, i.e. between giving priority to time targets or to cost targets; when calculating the net present value of an oilfield and using a fairly high rate of interest on capital invested, delays of production can be economically more harmful than even substantial cost increases over a few years.

(4)   The exception of rising oil prices in real terms can to a certain extent reduce the negative impact of the cost escalation on expected return on investment; in addition, to the extent that taxation contains capital allowances, the burden of the cost escalation is to a certain extent transferred from the concessionary companies to the government-landowner, and this can

indeed stimulate the concessionary companies to make expensive experiments with new methods of production.

Consequently, the escalation of costs can be seen as a function of increasing input requirements, as a function of increasing input costs, as a function of inadequate industrial organisation and management, as a result of a deliberate choice of a time-conscious strategy rather than a cost-conscious one, and finally as a result of a dynamic perspective on costs and prices together with unintended effects of government regulations, as well as inadequate government control. In both the UK and the Norwegian sector of the North Sea all these factors seem to be playing a role.

Among the cost factors, the rise in input costs is the easiest one to trace. In the immediate wake of the oil crisis, the North Sea was characterised by a high level of development activity and a strong demand for inputs.[60] In addition, there was inflation in most of the OECD countries, and particularly in the UK. Price rises were especially pronounced in steel, concrete and process plant. Manufactured equipment, which makes up about a third of the total cost of a platform, increased in price at the same rate as most capital goods during this period, i.e. 15 per cent in 1973, 30 per cent in 1974 and 20 per cent in 1975.[61]

Labour costs make up a high proportion, 30 to 35 per cent, of total platform costs. There has been no shortage of labour for the oil industry, in the UK or Norway, but in both countries the general wage level rose significantly in the mid-1970s. In addition, premiums have been introduced for workers in the oil industry because of the uncomfortable and hazardous working conditions, and the need to attract skilled personnel from other industries. According to one index, wages in UK concrete yards rose by 58 per cent between August 1973 and August 1975.[62] In Norway there have been similar increases.

The cost of several services has risen significantly, affecting the installation of platforms, pipelines and charges for supply boats, tugs, barges and the like. The high charges reflect the high capital intensity and risk of these services as well as the high demand for these services in the mid-1970s, which has made it possible for such service firms to operate as temporary monopolists.

The low productivity in UK industry perhaps explains part of the cost escalation, but for products delivered from Norway, where productivity is traditionally much higher and labour relations are much better, there has been similar escalation of costs.

It is difficult to determine the quantitative impact of the rise in input costs on the total escalation of costs. It is clear that the rise in input costs alone has hardly exceeded 30 per cent in any year in the 1970s, while total cost estimates have risen by 50 to 100 per cent a year in the same period. Consequently, external factors to the oil industry and to North Sea operations can only explain part of the cost escalation. The rest, and perhaps the bulk of the cost rise, is explained by internal factors.

As was pointed out earlier, the task of development, particularly in the northern part of the North Sea, is much more complex than development projects undertaken by the international oil industry elsewhere. For example, experience from the Gulf of Mexico has only limited relevance for the North Sea, and experience from the southern part of the North Sea has only limited relevance in its northern waters. As certain concepts of field development originating in more friendly offshore conditions, mainly the Gulf of Mexico, were applied with some success in the southern waters of the North Sea, they were to a large extent also attempted in the northern waters.[63]

However, the result was that technical problems and construction and installation problems were seriously underestimated. Many designs used in the Gulf of Mexico and the southern waters of the North Sea were scaled up and considered sufficient for the northern waters, where tougher weather conditions and greater depths create a qualitatively different environment. Not surprisingly, these designs were often inadequate, and had to be changed at great cost and with serious delays.

Granted, the task was not easy, because in the early 1970s little information was available on the environmental conditions in the northern waters and their impact on installations. At the time there was no institution providing a systematic collection of relevant data, and there were no UK or Norwegian construction codes for offshore installations. This lack of systematic information and construction codes has its background in a neglect of research and development efforts, and here the governments are mainly to blame. Only in 1974 were construction rules introduced in the UK and Norway for offshore installations, and even then the base of information was hardly sufficient.

The problems encountered necessitated design changes that were often passed on to the equipment producers late in the programme. These changes in production were both expensive and time-consuming. Only at a relatively late stage did the number of design changes stimulate the operating companies to review the whole concept of field development. This in turn led to new delays and more cost increases.

Until the cost escalation reached fairly alarming proportions in 1975–6, the UK and Norwegian governments seemed to have had an uncritical faith in the ability of the international oil industry to master the problems of development. This, perhaps, is the major reason why they neglected research and development. Information on geological and meteorological conditions is readily available, but information on problems of the installation and use of production systems requires a prolonged period of research and development. Ideally, this effort should have taken place before design and construction started, but in the North Sea the government effort in research and development was only stepped up when the design and production processes were fairly advanced.[64]

The operating companies not only drastically underestimated the size and the complexity of the development tasks in the northern waters, they were also slow in identifying the problem of cost escalation, and furthermore they have not been particularly efficient in solving either of these problems. This may seem surprising, as the operators are usually large international oil companies. For example, the operator of the costly Statfjord development is Mobil. The explanation for this lies at least in part in the industrial organisation of these companies.

Most of the large international oil companies have concentrated their skills in oil exploration, production, transportation, refining and marketing. Their goal has been to provide a fairly steady flow within a vertically integrated organisation. The development of a field is in most cases not a part of this integrated chain.[65] The reason is that the need to develop new fields fluctuates significantly over time, and thus so does the demand for the necessary equipment and expertise. It has been a fairly common practice for the large oil companies to contract out large parts of the effort of field development. In normal operating environments, such as the United States and the Middle East, the oil companies have had the expertise to supervise the work carried out by others. In addition, most of the projects in the traditional oil provinces do not differ dramatically in their requirements from their predecessors. This facilitates planning, implementation and control, and it also facilitates contracting out large construction projects. As a result, many of the large international oil companies have fairly limited experience in managing large complex construction projects. The usual practice is for the large oil company to engage separate designers, contractors and subcontractors to provide detailed designs, construction and assembly, and equipment delivery. This practice of subcontracting also means that the responsibility for development is diluted.

This structure of management is the main cause for the cost escalation in the northern waters of the North Sea. The major oil companies were in a hurry and did not realise the qualitative differences in development in the northern waters of the North Sea. Therefore they followed fairly uncritically the established practice of contracting out large parts of the task, in many cases through an independent managing agent, to a large number of contractors and subcontractors. This diluted responsibilities and produced badly defined structures of management, command and information. Not only did the large oil companies lack experience with these kinds of projects, but in many cases the contractors and subcontractors were also inexperienced.[66] Frequent changes in designs and specifications led operators to opt for cost reimbursable contracts, and this contributed further to the escalating costs.

When the project control at the level of management has been strengthened, this has usually led to time targets being more explicitly emphasised in relation to cost targets. To sum up, the project organisation in the northern part of the North Sea has often been characterised by inadequate channels of communication and unclear relationships of responsibility. Thus, in several critical cases, both the operating company and the major handling agent have managed the project in an inefficient way. An example is provided by the Statfjord field, mainly in the Norwegian sector, where the operator, Mobil, and the major handling agent, Brown and Root, are subject to heavy criticism by the Norwegian government and Statoil for inefficient management and low productivity.[67] However, this behaviour may not be entirely irrational from a private point of view.

In any industrial project, the relationship between time and costs has to be weighed. A private company will usually give a high priority to completing the project on time rather than keeping costs down. This is rational in order to maximise the return on capital invested, measured by the net present value of the cash flows and the internal rate of return, as private companies will usually have a fairly high rate of discount of the money. From this point of view, a delay in the start of production of a year or two can be more harmful than a significant increase in capital costs over a short period of time.[68] For a private international oil company, there are especially good reasons to act in this way. The rate of discount used in the oil industry is traditionally high, often 20 per cent. A heavy commitment in refining and marketing makes access to its own crude a key objective. North Sea oil, acquired through direct participation, is even after the recent cost escalation considerably less expensive and more profitable for the companies than oil bought from the state oil companies of the OPEC countries.[69] In addition, access to this crude gives an invaluable security of supply. Thus, a private international oil company has a natural propensity, given by its own premises, to give time a higher priority than cost control, at least within a wide range.

The pressure for speed is inherent in the development of most oilfields because of the companies' desire to borrow money for as short a time as possible. In the North Sea, this led the operating companies to reduce the period of planning and design, to start construction and installation at an early stage, and even to let the planning and design process overlap with the process of construction and installation instead of letting one succeed the other. Because construction started before planning was finished, revised plans often necessitated changing or rebuilding parts of the structure, causing both cost rises and delays.

From the point of view of governments, the relationship between time and costs appears in a different light.[70] This is especially true in the case of the Norwegian government, which was in a more comfortable economic position than the UK government. Given the UK balance of payments problem, the UK government had good reasons for pushing the development of new oilfields and giving time targets a high priority.

The Norwegian economy was in a much better position, and so the government could afford to show a certain amount of restraint and take time for good planning. This restraint is explicit in the Norwegian depletion policy, which will be studied in the next chapter, and in the overall licensing policy. It is a serious paradox in Norwegian oil policy that so far the restrictive and slow licensing policy has been accompanied by time pressure and forced work in the development of the individual fields. This essentially represents a conflict between macroeconomic considerations of the licensing government and the microeconomic concerns of operating private companies.

Given the size of the oil sector in relation to Norway's national economy, the cost increases and the delays, especially in the development of the Statfjord field, have led to serious macro-economic problems. The cost increases add to imports and the accumulation of foreign debts. They also mean that government revenues will be lower and later than anticipated. Norway by the middle of 1978 had a foreign debt of about $20,000 million, corresponding to about half of its gross national product. About one third of this foreign debt was from investment in the oil sector.

In Norway, an approach that emphasised thorough planning might have kept the cost escalation under control and increased the involvement of domestic firms, leading to a considerably lower import bill. As the balance of payments problem becomes less urgent for the UK government, similar considerations become increasingly relevant.

Table 3.11: Alternative Development Strategies and Cash Flows, Argyll Field

Alternative 1: Normal development, price of oil $12/b.
Alternative 2: Lead times respected, capital costs doubled.
Alternative 3: Cost range respected, lead time doubled.

Year	Capital Costs ($m)			Operating Costs			Output (10,000 b/d)			Cash Flow ($m)		
	1	2	3	1	2	3	1	2	3	1	2	3
1974	25	50	12	10	10	–	10	10	–	7	–18	–12
1975	–	–	13	40	40	10	30	30	10	92	92	19
1976	–	–	–	40	40	40	60	60	60	223	223	92
1977	–	–	–	40	40	40	60	60	60	223	223	223
1978	–	–	–	40	40	40	60	60	60	223	223	223
1979	–	–	–	40	40	40	60	60	60	223	223	223
1980	–	–	–	40	40	40	50	50	60	179	179	223
1981	–	–	–	40	40	40	50	50	50	179	179	179
1982	–	–	–	40	40	40	50	50	50	179	179	179
1983	–	–	–	–	–	40	–	–	50	–	–	179

Source: Author's own calculations on the basis of data provided by Wood McKenzie, *Political Economy,* and Mackay and Mackay, *The Political Economy of North Sea Oil,* pp. 46–7.

There is clearly a divergence of interests between governments and companies in the development of oilfields. The differences are illustrated in Tables 3.11 and 3.12 by comparing two variations of the anticipated development and cash flow of various oilfields. The hypothesis is that unexpected problems in field development can be overcome either by increased costs or by improved planning and longer lead times. In practice, the choice will not be as clear cut, but these two options reflect the two basic strategies of field development. In the following examples the effects of these two strategies will be analysed

Table 3.12: Alternative Development Strategies and Cash Flows, Forties
Field

Alternative 1: Normal development, price of oil $12/b.
Alternative 2: Lead times respected, capital costs doubled.
Alternative 3: Cost range respected, lead time doubled, with production starting in 1978
instead of 1975.

Year	Capital Costs ($m) 1	2	3	Operating Costs ($m) 1	2	3	Output (10,000 b/d) 1	2	3	Cash Flows ($m) 1	2	3
1972	150	300	75	—	—	—	—	—	—	−150	− 300	− 75
1973	300	600	75	—	—	—	—	—	—	−300	− 600	− 75
1974	600	1200	150	—	—	—	—	—	—	−600	−1200	−150
1975	300	600	150	30	30	—	50	50	—	−111	− 411	−150
1976	150	300	300	75	75	—	200	200	—	651	501	−300
1977	—	—	300	75	75	—	400	400	—	1677	1677	−300
1978	—	—	300	75	75	30	400	400	50	1677	1677	−111
1979	—	—	150	75	75	75	400	400	200	1677	1677	651
1980	—	—	—	75	75	75	360	360	400	1502	1502	1677
1981	—	—	—	75	75	75	324	324	400	1344	1344	1677
1982	—	—	—	75	75	75	292	292	400	1204	1204	1677
1983	—	—	—	75	75	75	262	262	360	1073	1073	1502
1984	—	—	—	75	75	75	236	236	324	959	959	1344
1985	—	—	—	75	75	75	213	213	292	937	937	1204
1986	—	—	—	75	75	75	191	191	262	762	762	1073
1987	—	—	—	75	75	75	172	172	236	678	678	959
1988	—	—	—	75	75	75	155	155	213	604	604	937
1989	—	—	—	75	75	75	139	139	191	534	534	762
1990	—	—	—	75	75	75	126	126	172	477	477	678
1991	—	—	—	75	75	75	113	113	155	420	420	604
1992	—	—	—	75	75	75	102	102	139	372	372	534
1993	—	—	—	75	75	75	92	92	126	328	328	477
1994	—	—	—	75	75	75	82	82	113	284	284	420
1995	—	—	—	—	—	75	—	—	102	—	—	372
1996	—	—	—	—	—	75	—	—	92	—	—	328
1997	—	—	—	—	—	75	—	—	82	—	—	284

Source: Author's own calculations on the basis of data provided by Wood Mackenzie,
Political Economy, and Mackay and Mackay, *The Political Economy of North Sea
Oil*, pp. 46–7.

in terms of a 100 per cent cost increase that avoids a time delay, and in
terms of a doubling of lead times with costs remaining in the range of
the original estimate.

The alternative strategies produce different financial results for the
two fields (see Tables 3.13 and 3.14).

Table 3.13: Alternative Development Strategies — Financial Results, Argyll Field

Indicator	Strategies		
	1	2	3
Payback, year	—	1975	1975
Internal rate of return	—	613%	348%
Net present value ($m)			
— at 20%	716	691	595
— at 15%	844	818	732
— at 10%	1,010	985	917
— at 5%	1,230	1,205	1,171

Source: Author's own calculations

Table 3.14: Alternative Development Strategies — Financial Results, Forties Field

Indicator	Strategies		
	1	2	3
Payback, year	1977	1979	1981
Internal rate of return	54%	32%	40%
Net present value ($m)			
— at 20%	2,494	1,437	1,316
— at 15%	3,783	2,636	2,383
— at 10%	5,868	4,622	4,302
— at 5%	9,437	8,177	8,003

Source: Author's own calculations.

In the case of the Argyll field, which in any case would have been most profitable, a delay of one year in production, as opposed to a cost increase of 100 per cent, would have given a considerably lower internal rate of return, 348 per cent instead of 613 per cent. At a discount rate of 20 per cent, the net present value of the cash flows would have been significantly affected, with a loss of about 13 per cent. At lower discount rates the difference in net present value is much smaller, for example at a discount rate of 5 per cent the difference is less than 3 per cent.

In the case of the Forties field, which is more representative, there is first of all a remarkable difference in the timing of the amortisation. According to the anticipated pattern of development, the initial investment would have been amortised after six years, in 1977. With the cost-intensive strategy amortisation is delayed two years, to 1979, and with the time-intensive strategy amortisation is delayed four years, until 1981. Because of the long production horizon of the field, the internal rate of return is more negatively affected by the cost-intensive strategy than by the time-intensive one; according to the initial pattern of development it is 54 per cent, whereas with the cost-intensive solution it drops to 32 per cent, and with the time-intensive solution it only drops to 40 per cent. At a discount rate of 20 per cent, the difference in net present value of the cash flows is in the order of 9 per cent; at lower rates of discount this difference declines, and at a discount rate of 5 per cent the net present value of the cash flows of the time-intensive solution is only 2 per cent less than that of the cost-intensive solution.

These examples show how it can be financially rational for a private oil company, relying either upon expensive borrowing or on a high rate of self-financing, to opt for the cost-intensive solution. The earlier amortisation and a significantly higher net present value at the rate of discount that is relevant will determine the firm's strategy. However, these examples also show how it is equally rational for a government that has easy access to capital and liquidity to see the trade-off between time and costs in a different perspective. For a government amortisation has less importance and the difference in net present value is hardly significant.

Apparently the cost escalation has a negative effect for both sides, reducing the return on private capital invested as well as the government revenue. This conclusion, however, is only valid in a static and short-term perspective. In a dynamic perspective, taking a longer time perspective into account, the cost escalation need not be a serious problem for the private concessionary companies. As previously indicated, the overall economics of the North Sea oilfields are generally extremely good at post-1974 prices, because development costs are small compared to the expected income over the lifetime of the fields. Consequently, the negative effects of even a substantial cost increase can be neutralised by a fairly modest price rise for oil. This is particularly true from the companies' point of view. Both UK and Norwegian oil taxation systems contain generous capital allowances, implying that capital costs can be written off several times. This, depending upon the characteristics of the field, means that to a certain extent the additional cost burden can be transferred from the concessionary companies to the government-landowner. Clearly, this provision does not stimulate cost-consciousness among the concessionary companies. Finally, given the high content of innovation and development in the projects in the northern part of the North Sea, the work carried out has to a certain extent the character of experimentation with a more general application for new offshore oil producing areas. Consequently, the tax system implies that the companies can liberally write off costs which are also research and development costs. Thus, for the companies, the North Sea can be a fairly inexpensive place for the oil companies to make experiments with new methods and systems of production.[71]

In recent years the cost escalation has been more acute in the Norwegian sector than in the UK sector, and particularly at the Statfjord project, compared to, for example, the Brent project. The two fields are close to each other, depths and weather conditions are similar, and the explanation for the difference in cost escalation should therefore essentially be with the management and the types of industrial organisation. The partial projects, Statfjord B and Brent D, are thought to be sufficiently similar to provide a basis for comparison.[72] Both projects make use of the Norwegian Condeep concrete production platform. With Brent D, completed at the end of 1977, the lead time was approximately three years. With Statfjord B, commenced in 1976, the lead time is anticipated to be six years. The cost of Brent D was $344 million (Nkr. 1,720 million). The estimated cost of Statfjord B is anticipated to be $2,060 million (Nkr. 10,300 million). Part of the difference can be explained by a general inflation, i.e. by input factors rising at 12 per cent yearly, by stricter Norwegian safety regulations, accounting for perhaps 20 per cent additional costs, and by more difficult conditions at the bottom of the sea, accounting for perhaps 10 per cent additional costs. Thus the cost reference for Brent D applied to current Norwegian conditions can be estimated at approximately $700 million (Nkr. 3,500 million), leaving a difference of $1,360 million (Nkr. 6,800 million), so, for practical purposes, the Norwegian Statfjord B platform is three times as expensive as the UK counterpart, Brent D.

The Brent D platform was for practical purposes delivered on time and to the price agreed upon. It is reported to be a success. The platform was built under the supervision of Shell, the operator, in Norway and it was equipped in Norwegian yards. The Statfjord B platform is being built under supervision by Mobil, the operator, and the practical details are handled by the agent, Brown and Root; building and equipment also take place in Norway. The number of hours accounted for by engineering designing, supervision, etc., is 7 to

12 times as high for Statfjord B as for Brent D. The operator, Mobil, decided without any bids that the main handling agent should be Brown and Root. The operator also has a fairly sovereign right to decide what companies are to get the different orders. The operator receives a certain percentage of the total costs as a compensation for the task, and there is compensation for handling the accounts and the expenses of the project. In this case the operator, Mobil, has a fairly small share in the field, 13.3 per cent. The Norwegian government, through Statoil, has a 44.4 per cent equity. However, when development was decided, in 1974, Statoil was two years old and considered, probably rightly, too inexperienced to handle the task. The Statfjord cost escalation has now reached proportions which could make it a major industrial scandal in Norway. There is a growing suspicion that the cost escalation contains at least an element of hidden transfer of economic rent. In March 1979 the Norwegian Parliament decided to set up a special commission to study cost escalation in the North Sea.

It is an easy proposition that neither private enterprise nor state participation has prevented cost escalation in the North Sea. Also, it would be unfair to place all the blame on the international private oil industry for cost escalation, even if it has been the operator in most cases. Essentially, the private international oil industry has acted according to its own premises, giving a high priority to time targets, using its established methods of project management, and perhaps seeking profits where they can be found. It is more surprising that the governments have not, or only belatedly, been fully aware of the propensity of private companies to be time-conscious rather than cost-conscious, and of the practical implications of the difference between private and public interests. It is remarkable that, for a long time at least, inadequate private planning and research and development have not been offset by a more comprehensive public planning at the level of field development. Also, government-sponsored research and development appear to have been fairly limited in relation to the tasks.[73] In some cases, inadequate public planning and preparation have led to unfortunate sudden changes in government directives and regulations. These have sometimes come at a fairly late stage in the project, with subsequent delays and cost increases. A classical example is offered by the Norwegian Petroleum Directorate, which at a late stage in the design process demanded a separate housing platform for Statfjord B, so that the whole concept had to be reviewed.[74]

Another not very flattering example is offered by the Statfjord A platform, which at first was built under great time pressure, and which later was delayed in port for a year before being towed out. Such examples indicate insufficient planning both with governments and the operators. Problems of this kind in many ways seem to have been more serious on the Norwegian than on the UK side. The explanation may be that Norway has a much less developed industrial base and less experience in handling problems of project management. The Norwegian tradition of administration also seems less capable than the UK one of handling this kind of problem, which can perhaps explain why the detailed control has been less satisfactory on the Norwegian side, leading to a higher cost escalation. Also, given the limited industrial base, Norwegian public authorities and industry could have felt less competent than foreign oil companies, and therefore had better not raise critical questions until a late stage. This kind of problem has evidently been much less present on the UK side.

Recent experience in the North Sea raises the question as to whether the international oil industry is technically and organisationally equipped to handle complex innovative tasks. The performance of Mobil at Statfjord may indicate that the company did not have sufficient experience, with the cost escalation being the cost of learning. Obviously, the international oil industry will learn from experience, but the problem is profound. The industry is likely to remain efficient in handling the classical operations – exploration, production, transportation, refining and marketing – but it is less likely to become specialist in complex innovative project management. Both British and Norwegian official studies of cost escalation stress inadequate planning, delegation of practical coordination to handling agents, and time pressure on the international oil industry.[75] Criticism is particularly severe in relation to project management where the operating company delegates both coordination and technical planning to one handling agent.[76] A recent Norwegian study recommends that the operating company keep responsibility for both these tasks.[77] Thus, the selection of operating companies should be based on their ability to manage complex innovative projects. Against this background, the main arguments for bringing in the international oil industry through the concessionary system may no longer be valid.

As an alternative, the international oil companies could be brought in through service contracts and entrepreneurial contracts, securing access to petroleum, but without controlling operations at the micro level. The other national and smaller companies could then have a chance of developing the industrial organisation required in the new areas. Financing at lower interest rates could induce them to make a

Figure 3.4: Typical Pattern of Project Organisation and Allocation of Management Responsibilities in the North Sea

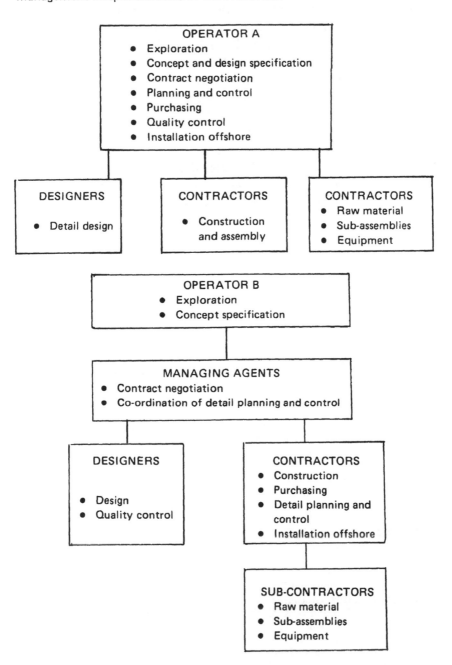

Source: *North Sea Costs Escalation Study,* Part II, p. 97.

trade-off between time targets and cost targets more acceptable to government. With hindsight, governments' somewhat uncritical belief in the organisational proficiency of the international oil industry may have been unavoidable, and simply a necessary stage in their learning process. The case of the Statfjord B platform, illustrates the point. The Norwegian Petroleum Directorate's demand for a separate housing platform, for safety reasons, came at a fairly late stage; it was opposed by the operator, Mobil, and by other forces in the Norwegian administration, who argued that it would involve an unnecessary cost increase with marginal impact on safety.[78] The Directorate was forced to drop its demand. Nevertheless, cost escalation has been serious on Statfjord B, and a recent study indicates that a separate housing platform would not have led to higher costs, but would have improved safety.[79] The real issue was pressure of time.

Notes

1. Keith Chapman, *North Sea Oil and Gas, a Geographical Perspective* (David and Charles, London, 1976), pp. 42 ff.

2. *Petroleumsundersøkelser nord for 62° N* (Ministry of Industry, Oslo, 1976), p. 33.

3. Peter Odell, 'The Economic Background to North Sea Oil and Gas Development' in Martin Seater and Ian Smart (eds.), *The Political Implications of North Sea Oil and Gas* (Universitetsforlaget, Oslo, 1975), pp. 51–80.

4. Ibid.

5. Carrol Wilson (ed.), *Energy: Global Prospects 1985 – 2000* (McGraw-Hill, New York, 1977), p. 126.

6. *Operations on the Norwegian Continental Shelf*, Report No. 30 to the Norwegian Storting (1973–4), (Ministry of Industry, Oslo, 1974), pp. 21 f.

7. *World Energy Outlook* (OECD, Paris, 1977), p. 47.

8. Chapman, *North Sea Oil and Gas*, p. 185.

9. Ibid., p. 211.

10. *World Energy Outlook*, p. 46.

11. Adrian Hamilton, *North Sea Impact* (International Institute for Economic Research, London, 1978), p. 64.

12. Ibid.

13. Jean-Marie Chevalier, *Le nouvel enjeu pétrolier* (Calmann-Lévy, Paris, 1973).

14. Ibid., pp. 49 f.

15. Michael Gibbons and Roger Voyer, *A Technology Assessment System* (Information Canada, Ottawa, 1974), pp. 39 ff.

16. D. I. Mackay and G. A. Mackay, *The Political Economy of North Sea Oil* (Martin Robertson, London, 1975).

17. Gibbons and Voyer, *A Technological Assessment System*, p. 42.

18. Pierre L. Bourgault, *Innovation and the Structure of Canadian Industry* (Information Canada, Ottawa, 1972), pp. 115 f.

19. Gibbons and Voyer, *A Technology Assessment System*, p. 47.

20. See, for example, Magne Østby, 'The Methanol Alternative for Producing Potential Gas Fields North of the 62nd Parallel on the Norwegian Continental Shelf' (Master of Science Report, Department of Petroleum Engineering, Stanford University, Stanford, 1975).

21. Bryan Cooper and T. F. Gaskell, *The Adventure of North Sea Oil* (Heinemann, London, 1976), pp. 118 f.

22. Ibid., p. 132.

23. *Ukontrollert utblåsing på Bravo 22. april 1977* (Oslo University Press, NOU, Oslo, 1977), p. 5.

24. *Petroleum Industry in Norwegian Society*, Parliamentary Report No. 25 (1973–4), (Ministry of Finance, Oslo, 1974), p. 30.

25. *Ukontrollert unblåsing på Bravo*, p. 11.

26. *Petroleum Industry in Norwegian Society*, pp. 32 f.

27. Ibid., p. 33.

28. Chapman, *North Sea Oil and Gas*, p. 102.

29. Ibid.

30. *Ilandføring av petroleum fra Ekofisk-området*, St.meld.nr. 1 (1972–3), pp. 56 f.

31. Chapman, *North Sea Oil and Gas*, p. 106.

32. Ibid., p. 112.

33. Ibid.

34. Ibid., pp. 116 f.

35. *Operations*, p. 14.

36. Mackay and Mackay, *The Political Economy of North Sea Oil*, p. 40.

37. Dankwart A. Rustow and John F. Mugno, *OPEC – Success and Prospects* (New York University Press, New York, 1976), p. 93.

38. Mackay and Mackay, *The Political Economy of North Sea Oil*, p. 40.

39. Chapman, *North Sea Oil and Gas*, p. 106.

40. Kenneth W. Dam, *Oil Resources* (University of Chicago Press, London, 1976).

41. Ibid., p. 98.

42. Ibid., p. 99.

43. Ibid.

44. Ibid., pp. 76–7.

45. *Operations*, p. 28.

46. Guy Arnold, *Britain's Oil* (Hamish Hamilton, London, 1973), pp. 37 f.

47. Mackay and Mackay, *The Political Economy of North Sea Oil*, p. 39.

48. *Act Relating to the Taxation of Submarine Petroleum Deposits*, (Ministry of Finance, Oslo, 1975), pp. 37 f.

49. Christopher Tugendhat and Adrian Hamilton, *Oil – the Biggest Business* (Eyre Methuen, London, 1975), p. 300.

50. Ibid.

51. Louis Turner, *Oil Companies in the International System* (George Allen and Unwin, London, 1978), pp. 200 ff.

52. Ewan Brown, 'Finance for the North Sea' in Saeter and Smart, *The Political Implications of North Sea Oil and Gas*, pp. 111–25.

53. H. E. Anonsen, *Financing the North Sea*, paper given at the Offshore North Sea Conference, Stavanger, Sept. 1976, p. 1.

54. Brown, 'Finance for the North Sea', p. 119.

55. *North Sea Costs Escalation Study*, Part II (Her Majesty's Stationery Office, London, 1977), pp. 44 f.

56. *Den Norske Stats Oljeselskap a.s.*, St.meld.nr. 33 (1977–8), (Ministry of Industry, Oslo, 1977), pp. 28 ff.

57. *North Sea Costs Escalation Study*, p. 52.

58. Ibid., p. 54.
59. Ibid., p. 54.
60. Ibid., Part I, p. 8.
61. Ibid., p. 9.
62. Ibid., p. 9.
63. Ibid., p. 10.
64. Ibid., Part II, p. 72.
65. Ibid., p. 94.
66. Ibid., p. 98.
67. *Den Norske Stats Oljeselskap a.s.*, St.meld.nr. 26 (1978–9) (Ministry of Oil and Energy, Oslo, 1978), pp. 8 f.
68. *North Sea Costs Escalation Study*, Part I, pp. 11 ff.
69. *Impact of Increased Taxation on Oil Exploration and Development in Alaska*, A Report to the Alaska State Legislature (Tanzer Economic Associates, Inc., New York, 1977), pp. 34 ff.
70. Stephen Marglin, 'The Social Rate of Discount and the Optimal Rate of Investment', *The Quarterly Journal of Economics*, vol. 77, no. 1 (1963), pp. 95–111.
71. Eilif Trondsen, 'Søkelys på konsesjonssystemet for oljeblokker', *Sosialøkonomen*, no. 3 (1979), pp. 4–6.
72. Brigt Hatlestad, 'Statfjordutbyggingen – en industriell katastrofe', *Norsk Oljerevy*, no. 2 (1979), pp. 6–7 and 22.
73. *North Sea Costs Escalation Study*, Part II, pp. 70 f.
74. *Den norske stats oljeselskap a.s.*, St.meld.nr. 33 (1977–8) (Ministry of Industry, Oslo, 1977), pp. 28 f.
75. *Kostnadsanalysen – Norsk kontinentalsokkel* (Ministry of Oil and Energy, Oslo, 1980), vol. 1, p. 24.
76. Ibid.
77. Ibid.
78. Bernt Eggen and Hakon Gundersen, 'Stilles i bero . . . ' in Bernt Eggen and Hakon Gundersen (eds.), *Nordsjøtragedien* (Pax, Oslo, 1980), pp. 103–43.
79. *Kostnadsanalysen – Norsk kontinentalsokkel*, vol. 2, pp. 243 ff.

4 GOVERNMENT STRATEGY

The development of government policies towards North Sea oil and towards oil companies is a dynamic process. Policy development is dynamic not only because of changing historical conditions before and after 1973–4, but also because policy-making is essentially a learning process. In this dynamic process the change in external circumstances was a shift or discontinuity, but the learning is continuous, as it started before the discontinuity and goes on afterwards.

Let us now look at how these dynamic elements of policy development have operated in the specific context of the North Sea. When first confronted with the possibility that there was oil and gas in the North Sea, neither the UK nor Norwegian governments had any direct experience with oil production. Their initial knowledge of the technical and commercial practices of the oil industry was scant. Furthermore, little was known of the possible effects of the oil industry on the economies and societies of the two countries, and the possible political pressures that might be created for the governments. Consequently, the governments could not give the oil companies a free rein, and they had to preserve their own freedom of action as much as possible in order to control the situation.

The rational policy solution was therefore to adopt a rather cautious approach, which meant keeping a good deal of control. This approach was in the self-interest of the bureaucracies. Any bureaucracy normally prefers stability and fears the unknown, and oil, although it brought bright economic prospects, was largely an unknown. The lack of knowledge was initially perhaps more accentuated in the Norwegian case, and was certainly an important factor in the government's decision on a cautious approach. As a result of their ignorance, both governments had to opt for a dualistic relationship with the oil companies: on the one hand, they were allies in order to explore and produce oil, and on the other hand, they were potential antagonists.

In developing their oil policies the two governments were operating within two distinct frameworks; one domestic and one international. In their national political systems and domestic economies, both the UK and Norway had, at least since 1945, strong traditions of government intervention in economic life. Through well organised trade unions and progressive taxation, both countries had experienced a considerable equalisation of incomes since 1945. Both countries had developed intensive systems of health care, welfare, pensions and other benefits. The public sector made up a comparatively high proportion of the gross national product in both countries, and there was a public industrial sector as well, relatively large in the UK and smaller in Norway. In both countries there was an established tradition, far older than 1945, that groups with economic and social grievances first addressed these to the state and the state was then supposed to intervene. Finally, in both countries economic and social policies after 1945 had to a large extent been elaborated by Labour parties. which were leftist social democratic parties, and which were in relatively close contact. Thus in both countries the government was at the centre of the economy, it was held responsible for the state of the economy, and it had at its disposal several means of intervention.

The economic situations in the two countries were rather different. Norway enjoyed a regular and relatively high rate of economic growth, while the UK had a low and irregular rate of economic growth that was related to its serious balance of payments problem. Consequently, in facing the potential bonanza of North Sea oil, there are rational reasons why the two governments acted differently. The British desperately needed an economic miracle while the Norwegians could do without one.

The second basic framework of UK and Norwegian North Sea oil is the international economy and international law. Both the UK and Norway have traditionally been very dependent on foreign trade and a smoothly working international financial system. For both countries the United States is an important trading partner, and both countries are US allies in NATO. Both countries have a long record of respecting international law, and both countries obviously have an interest in preventing international law from breaking down. In practical terms this meant that both governments respected their international agreements and commitments, even if they might appear unfavourable in the light of changing circumstances.

Given these two policy frameworks, domestic and international, the oil policies of the UK and Norway can be said to have two levels of rationality:

a political level that is essentially concerned with the government's ability to survive and maintain its freedom of action amid conflicting pressures;

an oil level, which is concerned with practical solutions to problems directly related to the oil industry.

In essence this means that the oil policies developed by the British
and Norwegians cannot be understood merely from the point of view
of oil itself, but involve a wider context.

At the political level there are at least six different considerations
concerning oil that should be identified:

(1)   the control of its macro-economic effects, such as the impact
       on economic growth, inflation, balance of payments,
       employment and the distribution of income;
(2)   the maximisation of public revenue from oil production;
(3)   the maximisation of spin-off effects from oil activities, i.e. to
       secure a place for national industry and to secure the largest
       possible part of the economic rent from oil for the national
       community;[1]
(4)   the control of regional effects, particularly the impact on
       local labour markets, and social effects;
(5)   control over the impact of oil production on the environment;
(6)   the integration of oil production into a national energy policy.

In Norway and the UK the failure to take these considerations into
account would sooner or later create political opposition to the oil
policy chosen, and could possibly endanger the stability of governments
and bureaucracies. The recent history of North Sea oil development
provides several examples of neglect of these political considerations.
In the UK the failure of the government to control the effect of the
oil industry on the balance of payments and the failure to maximise
public revenue led to a parliamentary inquiry, and a change in policy
towards heavier taxation of the oil companies.[2] Furthermore, the
initial failure to take regional effects into account has probably been
one of the factors behind Scottish nationalism.[3] In Norway, failure to
control regional and social effects effectively has provoked considerable
criticism. The failure to take safety and environmental considerations
seriously enough has led to a blow-out, followed by a report from a
commission of inquiry, blaming the accident essentially on administrative negligence.[4]

These political considerations are part of the realities within which
the oil industry operates, and they are as important as the purely
economic factors that are traditionally taken into account. These
political realities are superimposed on the more practical and

organisational questions that are relevant to oil exploration and production.

Depletion Policy

Depletion policy, the rate at which a finite resource is to be extracted, is a fundamental element of any oil policy. In many ways the depletion policy is the key policy element, as it heavily influences the bargaining position of the government in relation to the companies. A government opting for a high rate of extraction will necessarily be more exposed to the demands and needs of private companies controlling the relevant technology than a government opting for a low rate of extraction. In theory, the depletion policy is determined by three factors:[5]

> the expected rate of return;
> the changes in price that are expected over time;
> the variation of costs depending upon different rates of production.

For private oil companies, the expected rate of return will usually tend to be the same as the rate of return normally obtained on capital investment in private industry. For oil companies the determination of the rate of return may depend in part upon the full utilisation of integrated circuits through a steady access to crude oil. In general, in UK industry the rate of return has been around 15 per cent, while in Norway it has been less.

For governments the social rate of return or the benefits for the society as a whole will tend to be influenced by the absorptive capacity, or the greater the amount of unused productive factors in the economy, the higher the social rate of return. In other words, the greater the potential for using the revenue in the domestic economy, the higher the expected social rate of return. Correspondingly, if oil revenues must be invested abroad because of a limited capacity for absorption in the national economy, the relevant social rate of discount could be equivalent to the relatively modest return obtainable in the international financial markets.[6] The social rate of return considered by governments is basically dependent on macro-economic factors. This general trend is, however, modified by a few other considerations. Regional, social and environmental effects related to oil are determined by the rate of extraction. It can generally be assumed that these spin-off effects can be handled best when the national economy is prepared for the new market and its impact on the infrastructure and the environment. These considerations imply control and they theoretically tend to lower the social rate of return expected by governments.[7]

If negative effects result from oil extraction, this normally leads to government intervention to overcome the problems or compensate those affected adversely, which means additional public expenditures. This risk factor also lowers the social rate of return. Consequently, private and social rates of return may be quite different, with the social rate of return usually well below the private one.[8]

The future changes in the price of oil are as enigmatic to the British and the Norwegians as to anybody else in the Western world. Both are essentially marginal producers, in the sense that their volume of production makes up only a small fraction of the oil being traded in the world market. They cannot control prices by their level of output. However, in both countries there seems to be a fairly wide consensus that oil is likely to be more scarce in the latter part of this century and that prices are likely to rise quite considerably. When these price rises will come and what level they will reach remains, of course, an open question.

As mentioned earlier, North Sea oil development is now in a period of escalating costs and lead times. This is generally believed to be a normal pattern with a new economic activity. The stabilisation of costs and lead times can be assumed to be a function of the maturity of the activity. Maturity is gained through experience, and both time and extensiveness are probably elements of experience. Consequently, there could be a link between costs and the rate of production, apart from economies of scale. However, this cost factor seems so far to have been of rather limited importance to policy considerations.

Thus, in defining depletion policy in the North Sea, the relationship between the rate of return and the expected increase in the price of oil seems most important. Because the social rate of return tends to be lower than the private one, conflicts can arise between private and public interests in matters of depletion policy. If, for example, the expected rate of increase of the price of oil is between the private and the social rate of return, it is rational for private companies to accelerate production, but it is rational for governments to keep oil in the ground.[9]

So far the UK and the Norwegian governments have opted for remarkably different depletion policies. In the UK both Conservative and Labour governments have advocated a high rate of extraction, whereas Norway has quite consistently opted for a moderate rate of production. UK oil production is likely to reach 100 million tonnes, about the domestic level of consumption, around 1980, and it is expected to increase beyond that. In Norway, licensing has been quite restrictive and oil production is likely to be around 40–45 million tonnes by 1980. A major difference is that, while the UK will probably be a net exporter of oil for a period of 10–15 years, Norway is likely to remain in this situation at least well into the next century. It can of course be argued that the UK's normal production of 120 million tonnes is not particularly high for an industrial country of 56 million people. Furthermore, Norwegian production of 90 million tonnes of oil equivalents, which in Norway has been set as a moderate level, is not moderate for a country of 4 million people. However, this does not alter the basic difference in the perceived time horizon for oil production in the two countries.

In the UK the government seems to have come to the conclusion that there exists a relatively high social rate of return, which is not very surprising given the country's poor economic and financial situation. In Norway, on the other hand, the government seems to perceive a rather low social rate of return, emphasising the limited absorptive capacity of the economy.[10] Consequently, macro-economic considerations appear to be decisive for depletion policy in both countries.

In the UK, the trade-off between a high rate of production and the possible disruption it might cause seems to have been quite clear. The government seems to have been fully conscious of the possible negative side-effects for industrial, regional and social development.[11] In Norway, there was less of a dilemma, as considerations of industrial spin-off, regional, social and environmental effects essentially pulled in the same direction as the macro-economic considerations. This also explains why the negative aspects of oil development seem to have been given a relatively greater significance in Norway than in the UK.[12]

A psychological factor could also have contributed to the differences in depletion policies. The UK could have some problems identifying its interests with those of OPEC, because of its colonial past, its traditional position as a large importer of oil, and because it is the home country for two large international oil companies. Norway, by contrast, has a natural tendency to identify its interests with those of OPEC, given its situation on the periphery of Europe and its past experience of foreign domination.[13] The community of interests between Norway and the traditional oil producing countries in matters of production and prices has been explicitly stated by the Norwegian government.[14] However, such considerations have probably played a minor role. Essentially depletion policy seems to have been determined by macro-economic considerations.

Recently public opinion in the UK has been coming to favour a lower rate of extraction, stretching out the time during which the UK is an important producer of oil.[15] Evidently negative effects in matters of industrial, regional and social development are at the root of this change in attitude. New legislation from 1974 and 1975 allows the UK government to control the rate of production in existing licensees, but the government has also declared that this right will eventually be exercised most carefully, and in any case not until 1982.[16]

The disadvantages often associated with the administrative system are that it relies too much upon bureaucratic judgement rather than market forces. As a result, it does not gain the maximum part of the economic rent for the government, and it does not encourage efficiency among companies as well as the auction system does.

Table 4.1: Allocation Systems Compared

	Advantages	Drawbacks
Auction system	Secures large part of economic rent for the government Encourages efficiency among companies	Can defer exploration of less attractive areas Tends to exclude companies with smaller technical and financial resources Generally gives governments little direct control
Administrative system	Secures direct control for governments Encourages competition after allocation is done Can encourage exploration of less attractive areas Can secure a growth potential for companies with smaller resources	Can give a large part of economic rent to companies Does not encourage efficiency among companies Relies upon bureaucratic judgement

It should be noted that although the auction system emphasises gaining a maximum share of economic rent and the administrative system emphasises government control, the two are not mutually exclusive. Theoretically, the auction system can be supplemented by specified working programmes for each area, in order to meet government desires in exploration and development.[17] Also, the administrative system can be supplemented by special taxation systems designed to capture a larger part of the economic rent.

From the outset both the UK and Norway have opted for the system of administrative allocation. The reason for this choice is that it allows governments to discriminate between various applicants.[18] Both governments realised that they would be very dependent on American companies for a long time. As a result, the question of governmental control was naturally given high priority. Administrative allocation also guaranteed that domestic interests would get more advantages than they might have achieved under the auction system.[19]

The governments had very specific goals for the oil companies operating in the North Sea. They wanted the areas allocated to be explored quickly so that more could be known about the resource base. Administrative allocation explicitly allows governments to impose working programmes, and it also gave those applicants with the most comprehensive working programmes the best chances of getting a licence.[20] The governments generally believed that in an unknown area with high costs of production, such as the North Sea, the auction system might not lead to thorough exploration of the less promising areas. Another important goal for both governments was to have companies use domestic goods and services, in order to maximise the spin-off effects of oil production. The administrative system is more effective at this than auctioning.

So far, administrative allocation seems to have worked quite successfully in UK and Norwegian waters. Most notable has been the rate of exploration and the subsequent discoveries of petroleum. The system of administrative allocation was severely criticised in the UK for securing only a small part of the economic rent for the government.[21] However, this drawback is mitigated by successes in exploration and the growth of domestic companies. By contrast, Denmark, which in 1963 granted exclusive rights to a single consortium after an auction, has had minimal exploratory activity and little development of domestic competence.[22] The system of administrative allocation relies upon bureaucratic judgement, which probably meant some errors on licensing decisions in both countries. However, the total outcome so far is acceptable, and it seems reasonable to assume that greater reliance on auctioning and market forces might have produced less satisfactory results.

Initially, both the UK and Norwegian governments were facing the risk that oil companies with access to low-cost crude oil from the

Middle East might prefer to postpone development, keeping the higher-cost North Sea oil 'in the bank' for a long period of time. Government regulations therefore had to encourage companies to make an exploratory effort and produce from their discoveries.

Directly related to the problem of the licensing method was the licensing model. In principle, the choice was between two basic models:

a centralised model, in which exploration and production rights are granted for a large area to one company, or group of companies; a decentralised model, in which exploration and production rights are allocated in relatively small areas and in small numbers to groups of companies (see Table 4.2, p. 118).

The centralised model is used on the continental shelf in Eastern Canada and in the Danish sector of the North Sea. The decentralised model is used in the UK and Norwegian sectors of the North Sea.

The purpose of the centralised model is to give the licensee a lot of freedom in exploration and development. Normally, there is no specified working programme, and licensees often only have to use a certain sum for exploratory purposes. This facilitates planning in large areas as geological information is centralised. The centralisation of information in one company also creates the risk of one-sidedness in the treatment of geological data. From a government's point of view, the centralised model gives relatively little flexibility and little direct control.

By contrast, the decentralised model gives the government more control of exploration and development at the expense of the licensee. Detailed working programmes are usually specified in licences. This model does not facilitate company planning, as geological information is spread out among a number of companies. This decentralisation of information is advantageous in that the geological data are evaluated by several different experts from different companies. Governments have a lot of flexibility because the level of activity can be adjusted continually by the number of allocations. It also provides direct control, as licensing terms can be adjusted according to specific circumstances.

The main disadvantage of the centralised model is that the government has few chances to influence the behaviour of the licensee after the licence has been granted and thus cannot press for quick exploration. Another disadvantage is the possibility of errors that can result from the interpretation of geological data by one group as

Table 4.2: Licensing Models Compared

	Centralised Model	Decentralised Model
Flexibility	Low	High
Government control	Low	High
Treatment of information	Centralised, one-sided	Spread out, pluralist
Exploration	Slow, not very detailed	Quick, detailed
Planning of development of large areas	Easy	Complex

opposed to many. Theoretically, these disadvantages can be overcome if the government reserves the right to change the licensing terms after a given number of years, and obliges the licensee to share geological data with others. The main disadvantage of the decentralised model is that information is spread out and thus planning over large areas is a complex process, involving several companies. This drawback can be overcome by obliging companies to share all geological data with one central institution, preferably a government agency, and by instituting a system of state participation in all licences.

Historically, the centralised model has been dominant, particularly in traditional oil producing areas such as the Middle East. For a long time the practice was that the licensee got the exclusive right to a vast area with few, if any, obligations. As a result, the licensee gained a very strong position which usually created friction and ultimately the withdrawal of the concessions. Since 1945 most countries have opted for relatively smaller areas of concession. The decisions by Denmark in 1963 and by Canada in 1964 to award large areas to single licensees are clearly not in accord with this historical trend.

Relevant examples are the D'Arcy concession in Iran, granted in 1901 originally for an area of 480,000 square miles, the IPC concession in Iraq, covering by 1938 an area of 170,000 square miles, and the Aramco concession in Saudi Arabia, covering an area of 496,000 square miles.[23]

Given their preference for the decentralised model, the UK and Norwegian governments had to decide on the exact size of licence areas. The 'block principle', which is particularly widespread and was chosen by the UK and Norway, involves breaking a licence area into rectangular units that have borders defined in terms of latitude and longitude.

In the UK the size of blocks was determined in April 1964, at approximately 100 square miles per block. A year later, Norway opted for a block size of 500 square kilometres, or approximately 200 square miles. Several companies had approached the Norwegian government asking that the size of blocks be set at 1,000 square kilometres because less was known about the Norwegian sector, and it was less attractive from the point of view of petroleum geology. The companies consequently needed more freedom of action.[24]

Thus, from the outset, the UK opted for a licensing model and a size of area that could be considered strict for the private companies. The UK's goal was clearly to incite the maximum in exploratory activity. Norway was somewhat less strict in the size of areas. In both cases, provisions of relinquishment meant that the government would have large areas available again within a few years' time. Relinquishment involves the surrender of part of a licence area to the government after a given period of time, so that the licensee keeps only part of the original area.

Relinquishment can play a key role in a government's oil strategy, and it can have advantages for both governments and companies. The basic justification is the fact that exploration is expensive and risky, and only after extensive drilling is it possible to get an overview of the oil reserves of an area. Relinquishment implies that a government can initially license a relatively large area, knowing that part of it will be surrendered after exploration by private companies. In this way, relinquishment encourages private licensees to make a substantial exploratory effort in order to decide which part of the licensed area to surrender. The government will in time get back not only part of the area, but also more information about it.[25] This allows a government to justify licensing a larger area and at the same time review its oil policy on the areas returned by licensees.[26] By the same token, relinquishment allows private companies to get an initial licence for larger areas than otherwise would have been available, and after a given period of time abandon areas judged to be of less interest to them.[27] The principle of relinquishment, in order to be an effective incentive for exploration, must let the licensee decide which parts of a licensed area to surrender. However, current practice limits this choice through rules concerning the shape of the part surrendered. In most cases details are worked out in negotiations between companies and the government.

In the UK, the Petroleum (Production) Act of 1934 provided for the relinquishment of licensed areas. Licences were normally granted for 6 years, with the possibility of a renewal for 40 more years. At the time of renewal, at least half of the area licensed had to be surrendered to the government.[28] This clause was retained in oil policy for the North Sea.

In Norway, licence terms originally provided for a surrender of 25 per cent of the licensed area after 6 years, and of another 25 per cent after 9 years, with the total licensing period being 46 years, as in the UK. In 1972 Norwegian licensing conditions were changed, providing for a surrender of 50 per cent after 6 years, as in the UK system. The Norwegian system was also stiffened on another point. Previously, relinquishment concerned the entire area granted to a licensee. From 1972 on licensees have had to surrender half of each block granted.[29]

The length of time of concessions also presented both governments with problems. Historically, concessions to oil companies were granted for a long time. In Kuwait the original 1934 concession to the Kuwait Oil Company was granted for 92 years. The D'Arcy concession in Iran was originally granted for 60 years.[30] In Iraq the concession granted in 1925 originally had a duration of 75 years, and the one concluded in Saudi Arabia in 1933 was granted for 60 years. These long time periods were usually thought to be rather unsatisfactory for the producing countries.

In designing their concessionary regimes, the UK and Norwegian governments bore in mind that production costs in the North Sea were much less favourable than in the Middle East, and that, given the block system, the governments would keep a high degree of control. Consequently, a concessionary period of 46 years was not thought to be excessively generous. Also, it was feared that a short period might lead to an irresponsible depletion policy, with companies wanting to extract the maximum amount of oil in a relatively short period of time.[31] In 1972, when Norwegian oil policy was hardened, the duration was reduced to 36 years.

State Participation

UK and Norwegian oil policies use state participation to increase government revenues and exercise decisive government influence on new licences. State participation, as it is currently practised, is relatively new in both countries, and represents a significant change in the relationship between governments and oil companies.

In the period between 1945 and the early 1970s there was a gradual evolution towards greater governmental control of the oil industry in

the traditional oil producing countries.[32] The pre-1939 pattern had not
only been dominated by licensing models that were extremely
favourable to companies, including large areas and long periods of
time, but the companies were practically sovereign in disposing of their
licence areas.[33] After 1945 this pattern was seen as unacceptable by
many countries, and new forms of licensing emerged with state
participation as an essential component.

State participation could take several forms:

a carried-interest basis, in which the government participates from
the moment the licence is granted, or has an option to participate on
equal terms with other licensees if commercial quantities of
petroleum are discovered. No expenses are charged to the government until a discovery is declared as commercial. Alternatively, a
licence can also be granted to private companies when they are
linked with a state company on a joint-venture basis;
a service basis, which involves private companies acting as
contractors for a state company as the exclusive licensee;
a production-sharing basis, in which the private companies act as
contractors for the state company, and produced petroleum is
shared to an agreed formula.[34]

The purpose of state participation is threefold: the first is for the
government to secure the highest possible share of the earnings from
oil, apart from taxes, royalties etc.;[35] the second is to assure more
direct control of operations than is possible through licensing; the third
is to learn as much as possible about the oil industry through active
co-operation with private companies.

In the 1950s several of the traditional oil producing countries had
already opted for some kind of state participation in new oil
concessions.[36] Relevant examples are Iran and Saudi Arabia. By the
early 1970s it had become a cornerstone of oil policy in a large number
of producing countries. In the UK there was no direct state participation in the licences awarded in the first four rounds, but in some of the
blocks granted there was minority participation by public companies,
such as the Gas Council and the National Coal Board. With the fifth
licensing round in 1974, the principle of 51 per cent state participation
was established. This implied that the state was assured of a corresponding share of exploration and development costs. In Norway, a
provision for a minority state participation was included in the licences
awarded in 1969, and after 1972 the system of state participation on a

carried-interest basis was the general rule.

The British Petroleum and Submarine Pipelines Act of 1975 provided for a renegotiation of all previously granted licences, in view of the desire for 51 per cent state participation. However, the government had from the outset refused to impose state participation, and the problem therefore was to convince licensees to accept it.[37] Participation negotiations were carried out directly by a government committee that excluded the Department of Energy and BNOC, the state oil company that was to exercise the right of participation.

The negotiations had basic effects on ownership. First, the state got direct equity participation in smaller independent oil companies that needed additional capital because of rising development costs. The requirement of Ministerial consent for including new companies in a licence gave the government leverage in dealing with these capital-short companies. This procedure created a state majority stake in some holdings, but only minority participation in oilfields. Second, the state achieved indirect participation in cases where large international oil companies were involved by acting more or less as a banker.[38] At first the UK government proposed a solution akin to financial participation, with the government advancing 51 per cent of development capital and getting the right to 51 per cent of the oil produced, but with the companies being refunded their net post-tax revenue loss, with amortisation and interest on the initial participation payment being deducted.[39] The actual agreements seem to be that the government gets the right to buy up to 51 per cent of the oil produced at market prices.[40]

In Norway state participation was first introduced in 1969 as either a carried-interest or a profit-sharing arrangement, and state participation varied from 5 to 40 per cent. When a licence was transferred from one group of licensees to another in 1972, the government took the opportunity to increase its share from 26 to 40 per cent. In 1973 a licence was granted to a group composed of Mobil, Shell and Exxon, and for the first time state participation was set at 50 per cent. Participation was on a basis of carried-interest, with the state oil company having full rights in decision-making in the group, but with the private participants assuming the exploration costs. After exploration the state oil company had the option of deciding on participation, and eventually of taking over operating responsibilities ten years later.[41] Since the licensing round of 1974, Norwegian state participation has been at a level of 50 per cent or more.[42] Normally, the state oil company has an option of participating at a sliding rate of up to 85 per cent, depending on the size of a field.

Both the UK and Norwegian governments reserve the right to allocate entire licences to the state oil company. In Norway the government received parliamentary approval for giving the state oil company a 90 per cent share in a licence. This was made explicit in the announcement of the fifth round of licensing in the UK.[43]

The main difference between the UK and Norwegian forms of state participation is that the Norwegians explicitly practise the carried-interest system, while the British use a modified version of the production-sharing system. Consequently, the Norwegian system may appear somewhat more favourable to the government than the UK system, in that exploration costs are normally assumed by private companies alone, which eliminates risking public capital. However, given the fact that exploration is usually a small fraction of total development costs, the difference should not be exaggerated. Another important difference is that the Norwegian system of state participation is progressive, while the UK state share appears to be constant, regardless of the size of a field. This means that the Norwegian government is more likely to capture the extra surplus earned by large fields. Correspondingly, the Norwegian government is more likely to exercise a decisive influence in the development of large fields than the UK government.

In spite of its apparent success, the usefulness of state participation is sometimes questioned.[44] It is argued that the government could secure the additional economic rent just as well through taxation without having a state oil company. Correspondingly, it is argued that the control of the oil companies could just as well be organised through legislation and competition. These criticisms imply that state participation is a cumbersome way for a government to organise its fiscal and legal functions, indicating a lack of confidence in its fiscal and legal tools, and involving the creation of large new organisations that are costly and involve problems of steering. Furthermore, it could be argued that state companies do not have the same incentives for economic efficiency that private companies have. Consequently, it seems that participation could involve an economic loss to the society and the government. In the past, oil companies have been difficult to control from outside, whether they are private or public, and there is the risk that a state oil company might side with private oil companies against its own government. All of these points suggest that state participation could complicate the political control of the oil industry rather than enhance it.

These arguments are secondary to the way state participation is handled in practice. The essential argument in favour of state participation in the UK and Norway is that the governments found they had insufficient control over the private oil industry.[45] In both the UK and Norway state participation was justified as much by the need for control as by the need for revenue. In the UK and Norway the government has an acknowledged responsibility for safeguarding the important economic interests of the nation.[46] Thus government interest and objectives are not confined to earning, but also include the manner in which the resources are utilised.[47] Given the magnitude of the oil activities and their broad effects on society, direct and active participation was thought to provide the insight necessary to plan and control events.

In any case, unless the emphases on control and learning are mere excuses for a government participation that also increases the total government take, it is evident that the interests of government in a modern democracy in relation to the exploitation of a new resource are much wider and more complex than that of capturing the largest possible part of the economic rent. Control and learning, as a condition for control and planning, seem to be at least as important, and in many cases it seems that economic and financial objectives get a lower priority than the political objectives of control and learning. It can be argued, of course, that control and learning in the long run are essential conditions for maximising the government share of the economic surplus. In practice, however, it seems that state participation is not exercised purely, and perhaps not even principally, for this purpose.

State participation should be seen as a means to satisfy the varied political considerations that superimpose themselves on government oil policy, particularly the need to maximise spin-off effects and control undesirable regional and social effects. Because of government ownership, state oil companies can more easily be ordered to perform according to political criteria than private or foreign oil companies. In this way, state participation is not only a matter of national pride, it also serves the national interest.

The recent cost escalation in the northern part of the North Sea puts direct state participation in a new perspective. It was argued in the preceding chapter that the cost escalation could be explained to a considerable extent by factors that are inherent in the private international oil industry, i.e. the priority given to time targets rather than cost targets and the tendency towards delegating complex construction tasks to agents, contractors and subcontractors. The new

Table 4.3: Evolution of UK Concessionary Policy

Licensing Round	First	Second	Third	Fourth	Fifth	Sixth
Year	1964	1965	1970	1971–2	1976–7	1978
Initial fee	£25/sq km	£25/sq km	£30/sq km	£45/sq km	£80/sq km £120/sq km	£80/sq km £120/sq km
Fee seventh year	£40/sq km	£40/sq km	£50/sq km	£50/sq km	£200/sq km	£200/sq km
Yearly escalation	£25	£25	£30	£30	£200	£200
Fee ceiling	£290/sq km	£290/sq km	£350/sq km	£350/sq km	£3,000/sq km	£3,000/sq km
Royalty	12.5%	12.5%	12.5%	12.5%	12.5%	12.5%
State participation	None	None	None	None	51%	From 51% and up
Relinquishment	50% after 6 years	50% after 6 years	50% after 6 years	50% after 6 years	50% after 7 years	50% after 7 years
Licence duration	46 years	46 years	46 years	46 years	37 years	37 years
Work programme	Negotiable	Negotiable	Negotiable	Negotiable	Negotiable	Negotiable
Control of production	None	None	None	None	Included	Included

Source: Various British Government publications.

Table 4.4: Evolution of Norwegian Concessionary Policy

Licensing Round	First	Second	Third	Fourth
Year	1965	1969	1974	1978
Initial area fee	Nkr.500/sq km	Nkr.500/sq km	Nkr.750/sq km	Nkr.750/sq km
Fee seventh year	Nkr.500/sq km	Nkr.500/sq km	Nkr.1,800/sq km	Nkr.1,800/sq km
Yearly fee,escalation	Nkr.500	Nkr.500	Nkr.2,350	Nkr.2,350
Fee ceiling	Nkr.5,000/sq km	Nkr.5,000/sq km	Nkr.30,000/sq km	Nkr.30,000/sq km
Royalty	10%	10%	8–16%	8–16%
State participation	None	Up to 36%	50–75%	from 50% up
Relinquishment	25% after 6 years 25% after 9 years	25% after 6 years 25% after 9 years	50% after 6 years	50% after 6 years
Licence duration	46 years	46 years	36 years	36 years
Work programme	Negotiable (2–3 wells)	Negotiable (2–3 wells)	Negotiable (3–5 wells)	Negotiable (3–5 wells)
Hold development clause	None	None	None	For a specified period
Other conditions	None	Encourage Norwegian partners	Place orders with Norwegian industry. Industrial, financial and technical capacity.	Create jobs in Norway. Place orders in Norway. Financial and technical capacity. Experience in project management.

Source: 'Trends in Norwegian Licence Policy', *Norsk Oljerevy*, no. 6 (1978), p. 116.

state oil companies of the UK and Norway could provide an organisation that is more resistant to these problems. First, given their access to public capital at a low rate of interest, state oil companies should evaluate the relationship between time targets and cost targets in a different way, and they may be less pressed to start oil production for reasons of cash flow and return on capital. Thus, they could be less concerned about speed and more inclined to prolonged careful planning before construction. Second, the new state oil companies will have the bulk of their activities in the UK and Norwegian sectors of the North Sea, where the task of developing new oil and gas fields will be a central one. Thus, both in terms of the geographical scope and technical focus of the activities, the new state oil companies are likely to differ significantly from the international oil companies. For these two reasons, the new state oil companies may well develop characteristics making them more suitable for performing the particular task of managing complex development projects under unusually hostile conditions.

When the decision to establish state oil companies was made in Norway in 1972 and in the UK in 1975, the extent and causes of cost escalation were not known. The decision to opt for direct state participation through fully developed state oil companies may turn out to be wiser than was realised at the time.

Notes

1. Petter Nore, *Six Myths of British Oil Policies* (Thames Polytechnic, London, 1976), p. 4.

2. *First Report from the Committee on Public Accounts* (HMSO, London, 1973).

3. D. I. Mackay and G. A. Mackay, *The Political Economy of North Sea Oil* (Martin Robertson, London, 1975), pp. 111 ff.

4. *Ukontrollert utblåsing på Bravo 22nd April 1977* (Oslo University Press, Oslo, 1977).

5. Mackay and Mackay, *The Political Economy of North Sea Oil*, p. 43.

6. Nore, *Six Myths*, p. 5.

7. Ibid., p. 4.

8. Stephen A. Marglin, 'The Social Rate of Discount and the Optimal Rate of Investment', *Quarterly Journal of Economics*, vol. 77, no. 1 (1963), pp. 95–111.

9. Nore, *Six Myths*, p. 5.

10. *Petroleum Industry in Norwegian Society*, Parliamentary Report No. 25 (1973–4), (Ministry of Finance, Oslo, 1974).

11. Mackay and Mackay, *The Political Economy of North Sea Oil*, p. 44; *First Report*, p. 14.

12. Patricia W. Birnie, 'The Legal Background to North Sea Oil and Gas Development' in Martin Saeter and Ian Smart (eds.), *The Political Implications*

of North Sea Oil and Gas (Universitetsforlaget, Oslo, 1975).

13. Dankwart A. Rustow and John F. Mugno, *OPEC Success and Prospects* (New York University Press, New York, 1976), p. 115.

14. *Petroleum Industry in Norwegian Society*, p. 13.

15. Kenneth W. Dam, *Oil Resources* (University of Chicago Press, London, 1976), p. 114.

16. Ibid., p. 115.

17. Mackay and Mackay, *The Policital Economy of North Sea Oil*, p. 30.

18. Louis Turner, 'State and Commercial Interests in North Sea Oil and Gas: Conflict and Correspondence' in Martin Saeter and Ian Smart (eds.), *The Political Implications of North Sea Oil and Gas* (Universitetsforlaget, Oslo, 1975), p. 95.

19. *First Report*, p. ix.

20. Mackay and Mackay, *The Political Implications of North Sea Oil*, p. 25.

21. *First Report*, pp. 1 ff.

22. Turner, 'State and Commercial Interests', p. 95.

23. Jens Evensen, *Oversikt over oljepolitiske spørsmål* (Ministry of Industry, Oslo, 1971).

24. Ibid., p. 18.

25. Dam, *Oil Resources*, p. 50.

26. Evensen, *Oversikt*, p. 20.

27. Ibid.

28. Dam, *Oil Resources*, p. 50.

29. Ibid., p. 62.

30. Evensen, *Oversikt*, p. 29.

31. Ibid., p. 33.

32. Ibid., p. 64.

33. Ibid.

34. *Operations on the Norwegian Continental Shelf*, Report No. 30 to the Norwegian Storting (1973–4) (Ministry of Industry, Oslo, 1974), p. 45.

35. Ibid.

36. Evensen, *Oversikt*, p. 64.

37. Dam, *Oil Resources*, p. 118.

38. Ibid.

39. Ibid., pp. 107 ff.

40. Ibid., pp. 116 f.

41. Ibid., p. 60.

42. *Operations*, p. 64.

43. *Development of Oil and Gas Resources of the United Kingdom 1977* (Department of Energy, London, 1977).

44. Dam, *Oil Resources*, pp. 137 ff.

45. *United Kingdom Offshore Oil and Gas Policy* (Department of Energy, London, 1974), p. 2.

46. Ibid., p. 3.

47. *Operations*, p. 4.

5 ORGANISATION AND CONTROL

The Organisational Problem

The introduction of the oil industry into the UK and Norwegian economies caused structural changes that had to be met with a positive response at the governmental level. The governments had to create a certain harmony between external tasks and internal organisation.[1] In practice, this meant organising a government administration responsible for the supervision of oil activities. Later, with state participation and ownership, the problem was the organisation of a state oil company. As oil activities in the North Sea grew, the UK and Norwegian governments were continually confronted with difficult choices. They could not satisfy all interests, and consequently made explicit political choices. This process has had a profound effect on the structure and functioning of government in relation to economic life, particularly in Norway, but also in the UK.

The basic problem was that governmental changes designed to cope with new problems related to oil could also reduce government cohesion and effectiveness. This dilemma is constantly present in the choice between using existing government agencies for new functions and creating new government agencies for these new tasks. The dilemma is particularly important in setting up a state oil company.

In modern capitalist democracies, such as the UK and Norway, government cannot be seen as a unitary organisation pursuing a clearly defined hierarchy of purposes.[2] Instead, these governments are complex systems pursuing multiple and contradictory purposes.[3] The goals are usually determined by the needs of the external clients of the various government agencies. The activities of the agencies are to a large extent oriented towards satisfying external needs and demands. This analysis emphasises the service functions of government agencies over the co-ordination and policy formulated by central government organisations. In a Western European context, at least, with stronger traditions of economic and social planning than in North America, and with governments that are responsible to parliaments, the impact of central initiatives should not be neglected, but this impact can be highly irregular, like the initiatives themselves. Consequently, government organisations are under constant cross-pressures, and the forces exerting the strongest pressures are the most likely to influence decisions. Thus, government can be seen as a system of tensions.[4] This applies both to the individual government organisations and to the relationship between government organisations. The organisations of government can be formally ranged according to two criteria: the hierarchy of competence and the field of functioning, generalised or specialised. In any government some organisations will have a higher formal competence than others, for example in matters of budgeting, where the Treasuries or Finance Ministries have the highest formal authority, in co-ordination and in presenting matters of dispute to the executive. Also, there is a dichotomy between organisations with generalised functions, such as justice, social affairs, environment, etc., whose competence covers the whole society and all economic life, and specialised agencies, such as departments of agriculture, industry, etc., whose competence covers a defined economic sector.

The behaviour of these organisations can be seen in the light of three theories:[5]

> as determined by rational decisions, clearly defined preferences, insight into alternatives and knowledge of methods to maximise expected utility;
> as determined by bargaining and conflict resolution between competing organisations with different interests, where no single organisation alone can force a decision, and where resources are decisive;
> as determined by responses to events already taking place, i.e. more or less passively adjusting to changing external circumstances.

These three theories are not mutually exclusive: they complement each other and can together give a comprehensive picture of the workings of government. They are all relevant in the behaviour of the UK and Norwegian governments in oil policy. It seems that the depletion policy was decided rather rationally, with defined preferences and insight into alternatives. In the next step, in elaborating concessionary policy, taxation policy and state participation, there was a bargaining and conflict resolution between governments and oil industry. Finally, in some matters considered to be of subordinate economic importance, it appears that relevant government organisations were not extensively consulted, and that they were forced to adjust by outside pressures.

On this background, it appears that the level of rationality was quite selective, and that the potential to develop and use the knowledge in the relevant system of organisations was only partly employed. This has to do with the fragmentation of the power structure, i.e. a power that is relatively decentralised and corresponds to a series of partial purposes or objectives, and partial government processes, obstructing the development of an overall understanding of the situation.[6] This lack of overall concept of solution, or policy, again has a negative effect on the ability to find adequate solutions to partial problems, as the links will remain obscure. The decentralisation of planning in government gives a high degree of flexibility in individual matters, but efficiency is limited by an inadequate overall planning and inadequate steering in the relationship between government organisations. This raises the question of how the power structure in government can be changed, for example in order to be better able to cope with new problems. A major problem is the rigid structure of government organisations, which impedes experiments and to a certain extent learning, and which, through a nominal hierarchy and formal division of tasks, tends to obscure the real workings of the system. Consequently, it can be of interest to look at some theories on how the relationship between government organisations works out in practice.

In the relationship between large organisations, influence can be seen as a function of information, of finance and of learning. Generally, the following may be assumed:

> the organisation that is best informed, on a given subject as well as on its rival organisation, is more likely to influence the final decision; consequently, the steering potential can be seen as a function of resources to collect and to process information;
> the organisation that has the best access to funds is more likely to resist outside influence; consequently, the steering potential can be seen as a function of self-financing;
> the organisation that has the largest variety of actions at its disposal, as tools for defence and control, in order to expose the rival organisation to disturbances and variations in signals has the greater chance to influence the other; consequently the steering potential can be seen as a function of inventiveness and of the ability to develop an overall picture of the situation.[7]

In this way, organisational, financial and human resources, together with practical experience, can be of greater importance for an actual steering relationship than the formal distribution of responsibility and decision-making.

On this background it should be evident that the organisational solution chosen by government to tackle a new problem can have quite substantial effects on the relationships of influence and steering in the whole system of government organisations. Two specific risks should be mentioned:

> the organisational solution is insufficiently specific, the problem is not getting its proper attention, is administratively neglected and develops harmful effects to society, and ultimately to the government as well;
> the organisational solution is excessively specific, the problem becomes isolated from the rest of the government context and a process of government reversal is commenced, producing ultimately an effect that can be diametrically opposed to original intentions.

The trade-off has both a practical administrative relevance and a political relevance, and it concerns concepts of government effectiveness.

There are several advantages in using established government agencies for new functions of administration and control, apart from keeping a check on the growth of public bureaucracy. An established agency has acquired administrative experience, runs relatively smoothly in most cases, knows practices in interpreting and enforcing rules in the rest of the administration, and normally has an elaborate network of formal and informal contacts with the rest of the administration. This increases chances that new problems are dealt with according to generally established practices. The drawbacks in using an established agency for new functions are that the special expertise required is not very highly concentrated, practices from other sectors of government are not always equally relevant to a new problem, and malfunctions related to the internal culture of established agencies can be particularly harmful in relation to new problems. This decreases the chances that a new problem will receive special attention. Allocating a new function to an existing agency can also imply that power is shifted, to the benefit of the agency receiving the new function. By widening its field of functioning, the information base is enlarged, budgetary freedom increased, and learning potential positively influenced. For example, in Norway it has been asserted that the function of supervising oil has increased the power of the Ministry of Industry in relation to the Ministry of Finance.[8]

The advantage in creating a new specialised government agency for a new problem is that it can be dealt with very specifically, with special expertise which is more highly concentrated. New practices can more easily develop, partly because a new agency has to rely a great deal on innovation and experiment. A new government agency has a good chance of developing its own culture in relation to its specific tasks. On the other hand, there are considerable drawbacks in establishing a new specialised agency for new functions of administration, apart from expanding public bureaucracy. New specialised agencies tend to lack administrative experience, their running is not always smooth, there is a considerable risk that practices in interpreting and enforcing rules in the rest of the administration are not very well known or deliberately ignored, and the network of formal and informal contacts with the rest of the administration tends to be rather poor. A new government agency can develop its own culture as rather distinct from that of the rest of the public administration; consequently, a new specialised government agency increases the chances that a new problem will get special attention, but it also increases the chances that a new problem will be dealt with differently from generally established practices.

This problem is not only one of technical administrative relevance, it also has political importance. Any government agency in its administrative and supervisory functions basically has to rely on information furnished by its clients, the subjects of its control. This again means that, except in a few matters that are highly politicised and subject to considerable political debate, creating strong pressures from outside, the policy of a government agency in practical matters is based upon premises given by its clients. The more complex the subject-matter of the agency's supervisory activities, the more difficult it is for the agency to control the information given by its clients.[9] The more well organised the clients, and the wider their contacts with the government agency, the greater their ability to define premises and ultimately to influence the policy of the supervisory agency. An established government agency with rather general functions of supervision is unlikely to be very well informed on detailed matters, but it is also unlikely to have exceedingly close contacts with any of the individual subjects of its supervision. A new government agency, with specialised functions of control, is likely to be well informed on detailed matters, but it is also likely to be in rather close contact with some of its subjects of control. Consequently, generalised agencies carry the risk of lenient control, whereas specialised agencies carry the risk of being strongly influenced by external interests.

Generally, business enterprises can be assumed to be better organised and have larger resources for collecting and processing information than government agencies. This is particularly true of large business enterprises, whether private or public. This gives a high degree of flexibility and potential selectivity in the feedback of information that they give to a supervisory government agency. Any relationship with government and any systematic relationship of control requires that the controlling system has a greater repertoire of measures and variations than the controlled system.[10] Consequently, resources for collecting and processing information are relevant for the relationship of control, and a serious discrepancy in this respect, to the detriment of the controlling system, carries the risk that the intended relationship with government will be reversed, so that the originally controlled system starts governing the originally controlling system. In most cases, the government agencies will have few possibilities for competing with business enterprises in collecting and processing information, and consequently in most cases they seem to be in a structurally inferior position in this respect. However, this inferior position can be overcome to a considerable extent by developing an overall picture of the total situation, i.e. of the whole social and political context within which the business enterprises that are the subjects of control operate.[11] Consequently, the behaviour of the business enterprises and the different measures proposed can be judged less in relation to details, where the government is in an inferior position, than in relation to an overall picture of the situation, where the government can be in a superior position. Thus, effective government requires a vision of the functions of the controlled systems in a wider context and a certain politicisation of the matter, where value preferences are explicitly pronounced. In this way, given a vision of the whole context and explicit value preferences, government can develop a repertoire of measures that are outside the self-interest of the controlled systems, and in most cases outside their imagination as well.[12] On the basis of this discussion at least two levels of effectiveness of government control should be distinguished:

a micro-effectiveness concerning the steering at a detailed level, according to specific and technical criteria;
a macro-effectiveness, concerning the steering at a social and political level, according to value preferences.

It can be argued that, without a certain macro-effectiveness, the micro-effectiveness is of limited value for a government supervisory agency, as the aim of the control is to have the subjects perform according to general criteria that necessarily involve explicit value preferences that are less likely to be eroded than specific technical criteria which can be subject to continuous discussion, revision or negligence.

This distinction between micro-effectiveness and macro-effectiveness is also of considerable relevance to the structure of government organisation in relation to a specific problem. Briefly, two concepts of the structure of government organisation can be distinguished:

> a horizontal structure, consisting of hierarchical layers of government agencies with generalised functions, so that each one has working contacts with several others, without a 'privileged' relationship;
> a vertical structure, consisting of functionally defined sectors of government, with agencies having specialised functions, and each one having working relationships with a few others, with many 'privileged' relationships.

Generally, a horizontal structure fosters macro-effectiveness in government, while a vertical structure produces micro-effectiveness. Correspondingly, a horizontal structure carries the risk of a low micro-effectiveness and a vertical structure carries the risk of a low macro-effectiveness.

When organising government to cope with a new problem, the desire for detailed insight and control has to be weighed against the risk of 'contamination' from outside. The argument that effective government requires an overall picture of the situation is also to a large extent an argument in favour of a horizontal structure of government. The lack of detailed insight can certainly have a negative effect upon micro-effectiveness, but this is worthless without success at the level of macro-effectiveness.

On the other hand, it can be argued that horizontal structures of government are particularly well suited for a state that is passively controlling economic life according to general criteria of welfare and justice. Horizontal structures are less effective for the state that more actively intervenes in economic life. A state that seeks to intervene actively in industry must have specific tools and detailed insight, and this can justify the creation of specialised government agencies for new tasks, even if there is a risk of control being reversed.

In the discussion of government organisation in relation to oil in the

UK and Norway these general points should be kept in mind. When first considering the pattern of government organisation in relation to oil, the UK and Norwegian governments had inferior information resources compared to the international oil industry. This made a good case for creating special government agencies in order to improve detailed knowledge and the potential for micro-effectiveness. On the other hand, there was also a case for keeping matters related to oil under the auspices of existing government agencies with more generalised functions, partly in order to extend the authority of these agencies and partly in order to avoid creating an agency that served as the advocate of the industry it was meant to control.

The Choice of Organisational Pattern

In many countries close contacts between government agencies and the oil industry have clearly not enhanced government effectiveness. For example, in France a parliamentary commission concluded some years ago that the relationship between the Ministry of Industry and the oil industry was one in which the oil industry 'colonised' the government agencies responsible for controlling it and to a certain extent made them into its 'embassies' within the French state.[13] The oil industry has traditionally been difficult to govern, and relations between oil companies and governments have often been areas of conflict.[14] This fact was well known long before the oil crisis, and perhaps led the UK and Norwegian governments to take rather cautious approaches to government organisation.

At the outset matters related to oil were incorporated in existing government departments – in the UK the key roles went to the Ministry of Power, and in Norway to the Ministry of Industry. The only organisational change was that new low-level subunits with small staffs were created to deal with oil matters.

In the UK oil matters were transferred to the newly created Ministry of Technology in 1969, and in 1970 to the Department of Trade and Industry. In 1974 a Department of Energy was created, dealing with matters related to oil as well as more general problems of energy policy. This was part of a major reorganisation of UK government in relation to oil, carried out by the Labour government in 1974. Another part of the reorganisation was the creation of a fully state-owned oil company, designed as a fully integrated oil company.[15]

In Norway, after oil was found in commercial quantities, a special commission recommended a reorganisation of government in the spring of 1971 with relatively minor changes at the level of the Ministry, and the delegation of practical administrative functions to a separate Directorate. It also proposed the creation of a state holding company to handle the government's commercial oil interests.[16] The reorganisation proposal was reviewed and revised by the new Labour Government. The result was a major reorganisation of the Ministry of Industry, creating a special department of oil and mining and the formation of a separate Directorate for practical matters and the establishment of a new state oil company that was to develop into a fully integrated one. By the mid-1970s oil matters had become sufficiently complex, unique and time-consuming to require a Minister and bureaucracy of their own. In January 1978 the new Ministry of Oil and Energy assumed its functions.

Both countries decided to create new entirely state-owned oil companies, despite the fact that there already were some semi-public companies in their oil industries. In the UK, British Petroleum has been at least 48 per cent government-owned since 1914. Norway's Norsk Hydro was 48 per cent government-owned until 1971, and since then there has been government majority ownership of the company. Norsk Hydro has been developing considerable competence with oil both in Norway and abroad. Neither government found that it could use a semi-public company for its national oil company, despite the fact that they could have used their share-holding majorities to impose their will upon the companies.[17] This was not done, partly because the two companies were found to be 'untouchable', and partly because serious problems could have been created by the private owners.

Until the 1970s state oil companies have mainly existed in developing oil producing countries like the OPEC countries, and in industrialised oil consuming countries with high levels of state ownership, such as France, Italy and Spain. The creation of state oil companies in the 1970s in industrialised oil producing countries, like Norway, the UK and Canada, accelerated a trend that has been visible throughout the period since 1945 and signified a change in industrial policy in developed capitalist nations. The justification for creating state oil companies was essentially threefold:[18] to increase public revenue; to improve government control; and to improve government insight. In the Canadian discussion of the establishment of a state oil company[19] the following arguments were made in its favour:[20]

(1) social benefits and national pride through national ownership

of part of the petroleum industry;
(2) better knowledge of the workings of the oil industry;
(3) guidelines for government in collecting economic rent;
(4) influence in setting the price of petroleum;
(5) greater emphasis on refining;
(6) more research and development;
(7) use of national personnel;
(8) use of national goods and services;
(9) stimulus to regional development;
(10) an instrument in relations with other countries.

Norwegian considerations were rather similar. In addition to the desire to increase public revenue, the need for control through state participation was heavily emphasised.[21] In the British debate the need for control and insight was also emphasised along with the economic implications.

The Canadians also put forward arguments against creating a state oil company:[22]

(1) the cost of a state oil company would be an expensive undertaking for government for a rather long initial period;
(2) the state oil company would have to diversify into other areas because of the country's limited oil resources;
(3) most of the attractive petroleum sites were already leased by private companies;
(4) the creation of a large state oil company might diminish the interest of international oil companies in the country;
(5) a state oil company would be less economically efficient than private companies;
(6) so far, the behaviour of foreign oil companies was considered relatively satisfactory;
(7) there could be conflict between a state oil company and government, with the state oil company advocating its industrial interests;
(8) a state oil company would have no competitive advantage over private firms.

In the UK the debate over a state oil company was relatively polarised. The Conservative opposition claimed that the creation of a state oil company would be another step towards socialism.[23] In Norway there was a consensus that the state should participate more actively in the oil industry. There was, however, a debate on the form of participation.

On several occasions a state holding company was proposed instead of an integrated state oil company. It was believed that there was greater flexibility and economic efficiency in a small holding company participating through joint ventures and subsidiaries.[24] It was thought that this might allow for the gradual development of a national oil company with time to acquire expertise and use resources in a flexible and efficient way.

Norway ultimately decided to create a fully integrated oil company in a few years. This may give the government more direct control over and a better understanding of the oil industry, but it is also possibly a rigid and costly solution. In Norway there is a fear of the state oil company playing a dominant role in government and acting as the agent of foreign oil interests.[25]

The pattern of government organisation for dealing with oil is now remarkably similar in the UK and Norway. Both countries have:

specialised Ministries for oil and energy matters;
fully state-owned oil companies, organised as vertically integrated oil companies, following the pattern of the major international oil companies;
semi-public oil companies (British Petroleum and Norsk Hydro);
generalised government agencies that handle separate aspects of related oil issues such as taxation, environmental affairs and social and labour affairs.

In both countries this pattern of organisation has essentially been set up by Labour governments that invoke the necessity of an active state presence. Other parties might have opted for different solutions. In the UK the policy of the Conservative Party was to link matters related to North Sea oil with more general issues of trade and industry. In Norway, another government might have linked oil matters with trade at the level of Ministerial control, and would possibly have opted for a state holding company. However, Statoil was established by the unanimous vote of the Norwegian Parliament.

The Operation of the System

The UK and Norway now have essentially vertical structures of government to deal with oil. Direct links of responsibility and communication exist between the Ministries and the state oil companies.

The responsibility for general supervision of operations is embodied within this vertical structure. In the UK the Marine Division of the Department of Energy and in Norway the Petroleum Directorate, an external body dependent upon the Ministry of Oil and Energy, are in the supervisory roles. This vertical structure can be called the primary government structure in relation to oil.

In addition, there is a secondary structure of government for dealing with matters related to oil. It consists of government agencies with generalised or horizontal functions that also have a responsibility for clearly defined issues that involve oil. In fiscal matters the Treasury Department is involved in the UK and in Norway the Finance Ministry and its subordinate organisation, the Central Tax Board, are responsible. In matters of pollution control and preparedness for emergencies the UK Department of the Environment and the Norwegian Ministry of Environment, and its subordinate organisation, the State Pollution Control Office, are in charge. In matters of general safety a large number of existing government agencies are responsible for separate details in both countries. In labour and social affairs the Department of Health and Social Security in the UK and the Ministry of Social Affairs in Norway are responsible. In regional matters local bodies have some responsibility in both countries. In the UK, at the level of Ministerial responsibility the Scottish Office should be mentioned, and so should Norway's Ministry of Communal Affairs.

Finally, in both countries a tertiary structure of government can be distinguished in oil matters. This consists of permanent or *ad hoc* public bodies and committees with an advisory or consultative function. In both the UK and Norway such advisory bodies, often created to deal with a particular problem, can have considerable influence on oil policy. Examples are the Committee on Public Accounts in the UK and the Special Committees on Petroleum Taxation in Norway.

Within the vertical, or primary, structure, tasks are at three levels.

The policy level, i.e. the formulation of national objectives in relation to oil activities, the elaboration of law proposals and regulation, determination of order of priority of the different government tasks, co-ordination and direction of research and development, dealing with deliveries to oil activities, use of national goods and services, engagement in downstream operations, licensing and negotiations with private companies, determination of prices and royalties, supervision of subordinate tasks.

The control level, i.e. to supervise oil activities in detail to ensure

that laws and regulations are respected, to follow up policy
decisions, to plan and co-ordinate and partly to execute technical
control, and to remain in contact with relevant scientific institutions,
as well as to keep the policy level constantly informed on oil
activities.

The executive level, i.e. to implement oil policy in practice, to take
care commercially of the state's economic interests related to oil,
to dispose of the petroleum taken by the government as royalties, to
represent the government in state participation, to market oil and
gas, and to plan, co-ordinate and partly execute operations on the
continental shelf.

The major difference in UK and Norwegian systems of control is that
in the UK both policy and control functions are embodied in the
Department of Energy, whereas in Norway the Ministry of Oil and
Energy is more exclusively devoted to functions of policy, with
functions of control being largely delegated to the Petroleum
Directorate, an external body dependent on the Ministry. The reason
for delegating functions of control in Norway was to ensure that a
rather independent body could exercise control and collect relevant
information relatively independently from the Ministry, and thus
relatively immune to political pressures. The intention was obviously
to ensure that laws and regulations be adhered to in a rather strict way,
and it was feared that a control function kept within the Ministry
might have been more exposed to political pressures. Consequently,
the board of the Petroleum Directorate is established in a way that
makes it rather independent, even from the Ministry. In the UK such
considerations do not seem to have been very important, as control
functions were kept within the Ministry. However, a side-effect of this
division of functions in Norway is also to keep a certain distance
between the control functions and the Ministerial responsibility, and
consequently the parliamentary responsibility of the Minister and the
government. When control functions are directly embodied in the
Ministry, the Minister, and possibly the whole Cabinet, can more
easily be kept responsible for malfunctions and accidents than when
control functions are delegated to an external agency.

In both countries the main responsibility and powers are embodied
in the vertically organised primary structures. The powers give the
organisations of the primary structure the freedom of action required
to make rational decisions, weigh preferences and have the necessary
insight and knowledge to maximise expected utility. This is in many

ways also a prerequisite for the development and implementation of a rational oil policy. However, the powers also enable the organisation of the primary structure to dominate oil policy, and in particular in most matters related to oil to dominate the organisations of the secondary structure.

It is typical that in both countries the government organisations of the secondary structure have little or no direct influence on the elaboration of oil policy. Consequently, their influence and responsibility in practical matters are limited. This is illustrated by the fact that the main responsibility for preventing accidents that can harm employees and cause marine pollution is not vested with government agencies for social affairs or environmental protection, but with organisations within the primary structure – in the UK the Marine Division of the Department of Energy and in Norway the Petroleum Directorate. Consequently, the organisations of the secondary structure tend to have little direct influence on events, and their actions often tend to be responses to events already taking place.

This impression is corroborated by the fragmentation of the secondary structure, which consists of a number of quite different organisations with rather different tasks, corresponding to clearly defined partial objectives, implying partial processes of government. This makes each of these organisations rather weak in relation to the primary structure, so that in most cases of disagreement they can be assumed to avoid a conflict. This fragmentation also impedes the overall understanding of the situation, and consequently the formation of coalitions in the secondary structure of government. Consequently, in most cases of bargaining or conflict resolution, where a conflict really occurs, resources are likely to favour the organisations of the primary structure, given their information, their economic importance and possibly their superior understanding of the overall situation. Only in cases of quite wide coalitions of the organisations of the secondary structure can the distribution of resources be supposed to be to the detriment of the organisations of the primary structure. For such wide coalitions to develop, rather serious malfunctions would have to occur, within the primary structure and between the primary structure and society, giving an extensive politicisation of matters related to oil, so that value judgements and preferences will become more relevant. Normally, disagreements between the primary structure and different parts of the secondary structure will be channelled to the Cabinet. In the absence of an extensive politicisation, the resources of the primary structure can be assumed to be decisive at this level as well, particularly given the economic importance of oil to the governments. With an extensive politicisation, things could be quite different, with the primary structure encountering a formidable alliance of interests from the secondary structure, and with explicit value preferences and judgements imposing themselves at the Cabinet level.

It is still too early to reach any definite conclusions as to how these new vertical structures will function, and to what extent they will increase government effectiveness. The vertical structures result from the desire of the two Labour governments to control and participate in the oil industry. Given their large resources, these vertical structures are likely to achieve a considerable amount of micro-effectiveness.

This leads to the question of the macro-effectiveness of the vertical structures. The risk of such a vertical structure dominating the political system is greater in Norway than in the UK. However, in Norway oil issues are fairly politicised and thus the society's values and preferences necessarily impose themselves on politicians. If this politicisation of oil matters persists, government macro-effectiveness could remain high.

However, in the light of recent experience, the vertical structure might be modified. In Norway accidents and reports of bad working conditions have made the dominance of the Petroleum Directorate over safety and working conditions more controversial. Some generalised agencies, such as the Labour Inspection Agency, have taken over some of these tasks. Competition between parts of the primary structure and the generalised agencies would probably have a beneficial effect on safety and working conditions. In any case the vertical structure should be judged according to its performance. If it works well, the dominance of the primary structure will persist. But if conditions deteriorate and accidents occur, oil matters will be politicised and the dominance of the primary structure may be reduced considerably.

The Norwegian Parliament decided to maintain in existence the Petroleum Council, an advisory body functioning since 1965, giving it wider responsibility in defining oil policy, against the explicit desire of the Ministry. Given these political realities and the prospects that an alliance of interests from the secondary and tertiary structures could be activated in case of an open conflict, it is evidently in the interest of the primary structure to avoid direct clashes of interest and to exercise any dominance as discretely as possible. Against this background, there is likely to be an interesting learning process in both the primary structure and the secondary and tertiary structures. The organisations of the primary structure will increasingly learn the types of feedback created by their actions, and their strength, and consequently be more able to evaluate their own power position. The organisations of the secondary structure will increasingly learn their strength in relation to the primary structure, and the potential for coalitions and bargaining. Thus, a cycle of choices is developing, where each participant is learning the responses to different options.[26] The way this cycle develops will to a large extent determine the actual control of the oil sector, and the degrees of government effectiveness at macro- and micro levels. It will also considerably affect the political systems of the UK and Norway.

Furthermore, the general politicisation of matters relating to oil is likely to have a decisive influence, not only on the further development of oil policy and the interpretation of laws and regulations relating to oil, but also on the trade-offs between oil interests and social, regional and environmental considerations.

State and State Oil Companies

As mentioned previously, the major justification for establishing state oil companies in the UK and Norway was that they would give governments a better control over oil activities, as they would provide direct instruments for implementing government oil policy, and they would increase government expertise and insight in matters relating to oil. These considerations have, at least implicitly, been decisive when organising close contacts between the state oil companies and their subordinate Ministries in the vertical structure of government. However, using a state oil company as a tool for government oil policy implies that the government has full control and the necessary insight into its operations. The expertise and insight arising from direct participation in oil activities first of all goes to the state oil company.[27] The Ministry gets the relevant information only in the second instance. In the process of exchanging information between the Ministry and the private operators in the oil activities, the state oil company has the role of intermediary. The intermediate position gives potential for selection, implying a potential for deliberately influencing the flows of information. First in the transfer of information from the oil activities to the Ministry, the sheer volume implies selection, but this selection can easily become a deliberate screening. Second, in the transfer of information from the Ministry to the oil activities, some practical interpretation and precision is often required, but this again can easily become a deliberate reorientation.

Thus, within a system of state participation, the state oil companies occupy a strategic role, which in many ways gives them the advantages of both government and business. In relation to the government, the operational position of the state oil companies gives them evident advantages in the matter of access to relevant information. Furthermore, they generally have greater resources to collect and process information than their superior Ministerial organisations possess. In this way state oil companies are in the same position as large private business organisations, with the ability to be selective in the feedback of information, and consequently to try to keep the Ministry less informed than themselves. But state oil companies have an important advantage in relation to large private business organisations as well. The government's responsibility in any case implies closer links of communication with nationalised than private companies, and if nationalised companies are intended to perform particular tasks for the government, as is the case with the state oil companies, the links of communication are likely to be particularly close. In this way nationalised companies in general, and state oil companies in particular, have an advantage over private companies, and certainly over private oil companies, by being able to communicate more easily with government, and they are also more able to give premisses of decision to the government and thus to influence government policy. This can be called a political gain of nationalisation. In addition, state oil companies usually have access to low-interest treasury funds, which can make them view capital costs as being lower than those of private oil companies.[28] This can imply a lower rate of discount, giving a wider freedom of action in commercial ventures that can also be used for long-term strategic purposes. Correspondingly, in relation to the other oil companies active in exploration and production in the country, the state oil company has the obvious advantage of representing the government, so that the opinion of the state oil company is not only the opinion of an oil company, but also to a considerable extent the opinion of the state. In this way a state oil company can use its strategic position to build up a position of power by influencing government policy and by substituting itself for government in relation to private oil companies. Given this background, there is a definite risk that the active use of a state oil company in a system of state participation can lead to a gradual reversal of the process of government, so that real decision-making is transferred after some time from Ministries to state oil companies, and the role of government is reduced to that of endorsing decisions already taken elsewhere. In France, a tightly regulated system of oil refining and distribution, with state oil companies in a key role, seems to have led to such a reversal of the government function, with the state oil companies being the leaders in the matter of cartelisation and non-competitive behaviour for the whole sector.[29] Such a development is not only harmful to the government's economic interests, it is also directly contrary to the principles of democracy and of the government's responsibility. However, the advantages for the state oil companies as large organisations in this kind of development, giving them a much better control of their environment and consequently greater external stability as it does, are such that it is unlikely that the temptation will be resisted in the long run, even in the UK and Norway.

Furthermore, it has been an explicit objective of government policy to create efficient and dynamic state oil companies. An important aspect of dynamism and efficiency is the desire for independence from outside interference, in this case from the government-owner. The experience with the French nationalised oil industry is a telling illustration. In the UK and Norway, British Petroleum and Norsk Hydro are examples of semi-public oil companies which, though formally controlled by the government, are in practice independent.

The question is therefore what the governments can do in order to reduce this risk of government reversal. First of all, for the government to compete with the state oil companies in the matter of information collection and processing seems rather futile, for reasons of cost and organisation. A constant reorganisation of the state oil companies would reduce the risk, but would also be very detrimental to efficiency. Detailed supervision could also impede efficiency, lead to even greater bureaucratisation and to closer contacts, which might ultimately cause a reversal of the steering.

The best solution seems to be at the level of single steering mechanisms. Two possibilities are relevant:

the governments keep financial control over the state oil companies, by controlling their budgets;
the governments integrate the steering of the state oil companies into a system of general economic and social planning, providing a clear vision of the role of the state oil companies in the wider economic and social context. The governments can thus establish goals that are outside the narrow interest of the state oil companies.

The UK and Norwegian state oil companies are very similar in their basic relations with their governments. In both cases the state oil companies are wholly owned by the governments, and they are subject to relatively stringent controls. In the UK BNOC does not dispose of its own earnings. All oil revenue from royalties and state participation in BNOC is channelled into a special Oil Account at the Treasury, along with area fees and other income related to oil.[30] From this account BNOC's expenditures are financed. On the one hand, this arrangement gives the government considerable financial leverage with the company, because it eliminates the company's potential for financial independence. On the other hand, this arrangement could channel all public oil revenue to BNOC, which does not then have to compete with other government agencies in the budgeting process, and can satisfy its needs for capital easily.[31] Nevertheless, the arrangement also gives the Treasury an opportunity to control BNOC expenditures, and consequently its operations.[32]

In Norway Statoil needs the Minister's approval in all matters of principle or of major financial or political importance. Statoil is also obliged to submit each year a comprehensive survey of operations, which includes plans for development, plans to co-operate with other companies, proposed budgets and so forth. Statoil's earnings are so far not being channelled directly to the Treasury, but as of 1979 they have been insignificant. When Statoil's profits improve, this question might be reconsidered.[33] So far, Statoil's operations have essentially been financed through the Norwegian government. The extensive use of direct external borrowing by Statoil could reduce government control. In addition, Statoil's structure could also create some problems of control. The company is clearly aiming at a complex corporate structure, with a main company and several subsidiaries. Statoil is already involved in transportation through Norpipe and in refining and petrochemicals. The Comptroller General, the main vehicle of parliamentary control, regularly examines the accounts of public companies such as Statoil, but he has a limited mandate for examining the accounts of subsidiaries.

The control mechanisms of the two governments emphasise different methods. UK control of BNOC emphasises finances, while Norwegian control of Statoil emphasises information. The channelling of BNOC's cash flow through the National Oil Account gives the Treasury an opportunity for detailed control of the oil company's operations. In Norway, the Finance Ministry cannot exercise any similar detailed control of Statoil's operations, because Statoil is, for budget purposes, an independent organisation. On the other hand, the communication of general information and plans by BNOC to the Department of Energy seems sporadic. Thus BNOC's policy may develop quite independently from that of the Department of Energy. It should be underlined that the Treasury, not the Department of Energy, has the financial control. This control could be reduced to a mere accounting role, eliminating control through financial steering, and thus adding to BNOC's independence. In sum, BNOC's relation to the policy-making and controlling organisations of the UK government is ambivalent. On the one hand, the financial control of the Treasury limits the company's independence, but the division of authority between the Treasury and the Department of Energy could give BNOC a great deal of independence. The financial control exercised by the Treasury could also reverse the government's position. Close contacts between BNOC and the Treasury give BNOC ample possibilities to influence the Treasury and government policy. Also, given the importance of oil for the UK economy, and in particular for public revenue, BNOC's demands could easily get high priority in the Treasury.[34]

In Norway, Statoil must annually submit financial and operational plans and data to the Ministry and Parliament covering a four-year period. In this case, control is in theory exercised through policy-making. In practice, Statoil furnishes a comprehensive survey of its main activities and most important projects, including data on investment, financing, development and downstream involvement. There is relatively little room for detailed financial control, as Statoil's budget is neither approved by the Ministry nor voted by Parliament. The detailed financial control that does exist is essentially technical and is carried out by the Comptroller General. The control exercised by the Ministry is concentrated on broader economic matters. In the writing of the annual report on Statoil's activities there is fairly close contact between the Ministry and Statoil, and in the reports published and presented to Parliament the remarks of the Ministry are included separately and are sometimes critical.

Thus, in the Norwegian case, the position of the state oil company in relation to policy-making and government supervision seems more clearly defined than in the case of the UK. In Norway control and responsibility are firmly entrenched in the Ministry of Oil and Energy. The Norwegian system creates close contacts between the state oil company and the Ministry, which gives Statoil ample opportunities to influence policy. Also, given the importance of oil to Norway, Statoil's opinions could have a strong influence on the Ministry.

However, in Norway, the ability of the state oil company to impose its point of view on the government is contingent on economic growth and public revenue. As long as there is a strong need for more oil production to bolster growth and public revenue, the state oil company will be in a favourable position. When a surplus is reached, for example when additional income must be invested abroad because the economy has reached its absorptive capacity, then the economic need for more oil production will decline dramatically, and the political might of the state oil companies will also wane.

Because of Norway's limited economic absorptive capacity, this surplus is likely to be reached in the early 1980s. Consequently, it seems that Statoil's real potential to be a strong force within the Norwegian government will be limited to a relatively short period of rapid expansion. However, on this point depletion policy will necessarily be decisive.

In organising new state oil companies, it seems that the UK and Norwegian governments were quite aware of the political risks involved. In both cases the state oil companies were not organised according to the traditional public enterprise pattern.[35] The solution preferred was to tie the state oil companies closer to their governments – in the case of BNOC by financial control, and in the case of Statoil by regular detailed reports. In the UK nationalised industries have generally been organised as public corporations with wide freedom of action.[36] In Norway, where the public industrial sector has until recently been very small, nationalised firms were set up like privately owned companies and expected to operate on strictly commercial principles.[37] In recent years there has been a clear trend in both countries to use nationalised enterprises more consciously as tools of industrial policy. This implies direct government intervention, which in many cases leads to conflict between government and public firms.[38] For example, the nationalised enterprises are in most cases extremely reluctant to communicate detailed economic and operational data and corporate plans to supervisory government agencies. Usually, it is rare to find a corporate plan of a nationalised enterprise that has been altered because of government scrutiny.[39] This shows that government commercial organisations tend to develop an identity and interests of their own. However, the lack of adequate control is increasingly felt by governments to be a serious problem.

It was argued in the preceding chapter that BNOC and Statoil might provide a type of industrial organisation that is better suited for complex, innovative development tasks than the large international oil companies. Access to public capital could make the state oil companies less anxious about quick development and maximising the cash flow and return on capital, and their restriction to national waters would make them specialists in the unique conditions found in the North Sea. The recent experience with cost escalation is ultimately a challenge to both the governments and the state oil companies. For governments the problem is to design a steering mechanism and organisational context that encourages the state oil companies to give cost targets a higher priority than time targets, and integrate the various elements of a project. The lesson here is that the organisation of the new state oil companies should be specialised and tailored to the demands of the North Sea and not adopted uncritically from the large international oil companies.

Recently, both BNOC and Statoil have received direct external financing. In June 1977 BNOC announced a loan of $825 million from a group composed mainly of American banks.[40] The operation was made apparently without any Treasury guarantee and without mortgaging the oil belonging to BNOC. In 1978 Statoil received a loan of Nkr. 2,000 million ($400 million) from a group of foreign banks.

In the case of the UK there may be a purely political explanation. With a Conservative government coming to power, BNOC might be in danger of being liquidated. The external loan gives foreign financial interests a claim on the company, and therefore might be seen as a guarantee for BNOC's survival under a Conservative government.

In the Norwegian case the explanation is financial. In 1978 the indebtedness of the Norwegian state was great and any more external borrowing might seem irresponsible. Direct borrowing by Statoil would bring capital into the country without directly increasing the public debt, even if the loan is guaranteed by the government.[41] In a report to Parliament, the Norwegian government explicitly stated that external borrowing by Statoil is due to this special situation.[42]

Despite the fact that the direct external borrowing by the state oil companies was justified by special political or financial circumstances, a precedent may have been set. The continued use of direct external financing might gradually contribute to the erosion of the government's financial leverage with the state oil company. Direct external financing is also likely to promote a separate sense of identity in the state oil companies which could reinforce interests that are different from those of the state and perhaps more similar to those of large private oil companies. Such a development[43] appears fairly remote, but increased external financing could pull the state oil companies in a direction that is possibly not envisaged presently by themselves, and not desired by the governments.

In order to ensure this kind of tailoring, governments should influence the planning process in the state oil companies at the early stage when goals are defined. Government views on the activities of the state oil companies must be made explicit and taken into account by the companies. This can be done through a comprehensive system of economic and industrial planning which is presently lacking in both the UK and Norway. It can also be done through detailed sectoral planning and a stiff licensing policy.

Steering Through Licensing

In the North Sea model licensing is a key tool for government policy. Licensing is used essentially for two different policy purposes: it is the main instrument for regulating the overall level of activity,[44] and it is also the major instrument for influencing a salient quantitative aspect of the activity — the distribution of tasks and opportunities between different companies. The reliance of the North Sea model upon administrative licensing rather than licensing by auctioning means, at least implicitly, giving a higher priority to political control, and to control of the micro-economic aspects, than to the capture of the economic rent.[45] The choice of the administrative method of allocation is consistent with this preference and, as has been argued before, with the potential for differentiated political feedback to the government in the UK and Norway in charge of oil activities. From the point of view of government control, the administrative method of allocation has at least three advantages: access to information, selectivity and flexibility.

The control of the overall level of activity has three aspects: the control of exploration, the control of development and the control of production. For the government to control the rate of exploration is apparently the most simple of the three. Provided the geology and government terms make the areas attractive from the point of view of private companies, the rate of licensing and the adjustment of working programmes permit the government, at least in theory, to influence the rate of exploration fairly directly. For example, if an even level of exploration activity is desired, to keep a steady demand for rigs, supply ships, etc., this can be achieved by regulating the rate of licensing and adjusting working programmes over time. However, as is evident, there needs to be no direct relationship between exploration and discoveries.

Discoveries tend to be made unevenly over time and the fields discovered tend to be of quite different character. Historically, the largest fields in an oil province have been found first, with discoveries thereafter tending to be gradually smaller. There are, however, important exceptions. Also, the rate of discoveries and their character can to a certain extent be influenced by licensing policy. In any case, gross additions to reserves in a given oil province tend to vary considerably from the rate of exploration. Another salient point is that with the provision of relinquishment, as in the North Sea model, important discoveries tend to be made, or made public, several years after initial exploration was commenced, and just before the deadline when the companies have to decide which parts of the areas allocated are to be rendered to the government. Against this background, the rate of development can vary considerably over time, depending on uneven additions to the reserves. Consequently, in order to keep an even rate of development, the rate of exploration may have to increase.

The most complicated aspect is the control of the rate of production. Economically this is linked to the fact that in frontier oil provinces, such as the North Sea and adjacent areas, where exploration and development move into gradually more difficult conditions, there seems to be a tendency to diminishing marginal returns on development projects because of rising costs, slow-downs and even reductions of volume. If this trend is broken, for example by technological innovations, there will be a tendency to sharp discontinuities. Politically, the steering of the level of production is linked to the difficulty of not developing fields that are found. Whenever a discovery is made, economic considerations easily indicate that development should be commenced instantly, and that production should commence as soon as possible. This can, in theory, be offset by government regulations permitting development or production to be deferred, at the discretion of the government, or indefinitely. In practice this may be more easily said than done. In the early licences granted, the governments had hardly any possibility of controlling directly the rate of production, except for technical conservation purposes, i.e. in order to have a rational production profile giving the best productivity of a field. In the UK, the Petroleum and Submarine Pipelines Act of 1975 entitled the government to control the rate of production directly. This also applies to fields under licences granted before 1975. In Norway no such retroactive measures have been taken, but the terms of all licences granted after 1974 contain clauses that the rate of production will be subject to direct government control. In the preparation of exploratory drilling off northern Norway, the Norwegian government has explicitly stated its right to demand a deferral of development or production.[46]

Evidently, such provisions can be expanded, and the Norwegian government appears to be envisaging provisions under which exploratory drilling in a given licensed block can only take place after results are known from exploration on adjacent blocks. The purpose is to provide for a gradual exploration of adjacent areas and to avoid sudden discoveries disrupting the planned level of activity.[47] However, private participants will usually have an interest in quick development and in production being advanced as much as possible to gain return on capital invested. Thus, both the UK and the Norwegian governments are legally enabled to decide the level and the timing of production of the different fields in an authoritative way. For the government, the advantages are the ability to control the overall level of output in order to avoid sharp fluctuations and, especially in the case of Norway, to avoid production ceilings being broken. It also enables the government to plan ahead, in terms of development strategies, transportation solutions and even marketing strategies. But the question is to what extent a deferral or limitation of production will be accepted by the private concessionaries. If they see keeping oil in the ground as security for future supplies and as an economic advantage in the light of anticipated future price rises, it may be accepted. If they see it as economically harmful and as harmful to their relations with the governments and markets of the consuming countries, it may not be accepted. Also, it is an open question to what extent decisions by the UK and Norwegian governments to limit or defer oil production will be readily accepted by other OECD governments, with prospects of OPEC supplies becoming tighter.

The administrative method of allocation is based on a bargaining process between government and companies. The final allocation is, on the one hand, subject to political criteria which are exogenous to the bargaining situation, such as the desire to give national companies a chance to participate on preferential terms. On the other hand, the final allocation is dependent on information submitted by contending companies on intended work programmes, etc.

There is thus a rational incentive for the companies to provide the government with information. The work programmes relative to the different areas are determined by negotiations between the government and the applicant companies, where it is in the government's interest to have a speedy and thorough exploration of the area. During the allocation negotiations the government gains access to information on the technical competence and financial resources of the companies, as

well as on the companies' view on the areas open for allocation, because
this information, which would probably not be available otherwise,
provides criteria for government preference and allocation.[48] In the
United States, the government has attempted to make the companies
deliver this kind of information, but so far without much success, since
with licensing through auctioning there is no incitement for the
companies to provide the government with salient information on
themselves or the prospective areas. On the contrary, to hand over
critical data to the government could lead the government as the seller
of allocations to demand an initially higher price. This access to
information is perhaps the key advantage of the administrative
allocation compared to auctioning. The access to information is a
necessary condition for using the two other advantages, selectivity and
flexibility. The selectivity of administrative allocation implies that the
government to a large extent can decide what companies are to get
access to what areas. In a system of licensing through auctioning there
would be a tendency for the richest companies to get the most
prospective areas. The flexibility of administrative allocation implies
that the government can fairly easily adjust its policy according to
changing circumstances, for example in the choice of companies.

As already pointed out, discoveries tend to be made unevenly, and a
large proportion of the total reserves of a given oil province tends to be
located within a few large structures, so the attraction of different areas
tends to vary considerably. From the government point of view, it can
be important to get a clear picture of the total reserves in the less
promising areas as well, whereas it can be in the interest of the
companies to concentrate exploration in the most promising areas. The
selectivity of administrative allocation is in this respect a major strategic
advantage for the government. Through administrative licensing the
government can issue licence areas of different prospectivity in the same
licensing round. By combining the less attractive areas with the more
promising ones, better terms can be obtained for the former than if
they were licensed alone.[49]

Thus a more balanced exploration of the continental shelf can be
obtained. Smaller fields, which alone would hardly be commercially
exploitable, can be combined and developed with more attractive
discoveries. Also, in many cases, large discoveries have been made in
areas that have been considered less attractive, and vice versa.

Historically, foreign international oil companies have tended to have
more technical competence and larger financial resources than domestic
UK and Norwegian oil companies. Consequently, the foreign inter-

national oil companies have had a comparative advantage in relation to the newer domestic oil companies. From the point of view of government strategy, and to the extent that the government sees it as important to give preferential treatment to domestic companies in order to improve their technical competence and financial situation, the selectivity of the administrative allocation also offers important advantages. Through administrative licensing, the government can issue licence areas of different prospects to different kinds of companies, combining stronger and weaker companies in the same group, and thus giving the latter better working conditions than if they were operating alone. In this way, smaller and newer companies can learn, acquire expertise and in the longer run improve their financial resources. This is also likely to have a positive effect upon competition. Finally, administrative licensing offers the government the possibility of choosing at its own discretion companies which not only have the necessary technical and financial resources, but which are also likely to be loyal partners, for example on the question of production control. Also, by administrative licensing, allocations can be used as a reward for satisfactory behaviour.

In addition, as mentioned earlier, administrative licensing enables government to impose a certain pluralism. This can be of great importance in the treatment of geological information and in the choice of methods of development. There is therefore a strong case for issuing licences to groups of companies rather than to single companies. Like other organisations, oil companies tend to develop different characters, with dissimilar strengths and weaknesses, and consequently dissimilar methods of treating geological data and choosing methods of development. In this perspective, company pluralism seems to be strongly in the government interest. Consequently, giving the state oil companies a monopoly does not appear to be in the government interest, as this would be the end of pluralism. On the contrary, as the system of state participation becomes generalised, there is perhaps an even stronger case for maintaining private oil companies in UK and Norwegian waters. The presence of other companies can also be a tool of government, as through participation of private companies standards are created against which state oil companies can be judged, and channels of information are created through which alternative points of view may reach the governments.

So far, this pluralism has been assured through conventional licensing, by which other companies obtain ownership of part of the oil. However, pluralism can also be assured through a system of service contracts. The main difference between service contracts and the usual licences is that the private company or companies do not obtain ownership of the oil found. The usual pattern is that a private company risks capital in exploring for oil and gets the right to a share of the oil, at a price agreed upon. Many service contracts do not differ significantly from joint-venture contracts with state participation.

To sum up, administrative licensing offers the government a set of policy instruments which at least in theory give the government the right to decide all salient aspects of the activities. However, this steering is to a certain extent indirect, producing a web of government regulations and provisions to offset the right of the concessionary companies to dispose of the oil and organise oil activities. To control the level of production is the most complex and difficult aspect of this, implying serious contradictions of policy both with the basic principle of licensing itself, and even with the aims of exploration and development policy.

Notes

1. Eric Rhenman, *Systemsamhället* (Aldus, Stockholm, 1975), p. 32.

2. *Politische Ökonomie des heutigen Monopolkapitalismus* (Dietz Verlag, Berlin (East), 1972), pp. 383 ff.

3. Frederick C. Mosher, 'Analytical Commentary' in Frederick C. Mosher (ed.), *Governmental Reorganization* (Bobbs-Merrill, New York, 1967), pp. 475–545.

4. Ibid., pp. 487 ff.

5. Johan P. Olsen, 'Choice in an Organized Anarchy' in Johan P. Olsen and James G. March (eds.), *Ambiguity and Choice in Organisations* (Oslo University Press, Oslo, 1976), pp. 82–139.

6. Rhenman, *Systemsamhället*, p. 85.

7. Eric Rhenman and Richard Norman, *Formulation of Goals and Measurement of Effectiveness in the Public Administration* (SIAR, Stockholm, 1975), pp. 39 f.

8. Jon Naustdalslid, 'Oljen og styringsproblema: Om politisk-administrative konsekvensar av aktuelle styringsformer', *Internasjonal Politikk*, no. 2B (1975), pp. 335–51.

9. Rhenman and Normann, *Formulation of Goals*, pp. 35 f.

10. Ibid., p. 37.

11. Ibid., p. 39.

12. Øystein Noreng, 'Staten, oljen og de politiske kanaler' in *Om Staten* (Pax, Oslo, 1978), pp. 136–76.

13. *Sur les Sociétés pétrolières opérant en France* (Union Générale d'Editeurs, Paris, 1975), pp. 224 f.

14. Jens Evensen, *Oversikt over oljepolitiske spørsmål* (Ministry of Industry, Oslo, 1971), p. 10.

15. Edward N. Krapels, *Controlling Oil, British Oil Policy and the British National Oil Corporation* (US Government Printing Office, Washington, DC, 1977), p. 18.

16. Jens Evensen, *Innstilling om organisasjon for statlige kontinentalsokkelsaker* (Ministry of Industry, Oslo, 1971), p. 5.

17. Petter Nore, *Six Myths of British Oil Policies* (Thames Polytechnic, London, 1973), p. 18.

18. Krapels, *Controlling Oil*, pp. 20 ff.

19. Ghislaine Cestre, *Petro-Canada: A National Oil Company in the Canadian Context* (US Government Printing Office, Washington, DC, 1977), pp. 18 f.

20. *An Energy Policy for Canada – Phase 1 Vol. 1* (Department of Energy, Mines and Resources, Ottawa, 1973), pp. 186 f.

21. *Petroleum Industry in Norwegian Society*, Parliamentary Report No. 25 (1973–4) (Ministry of Finance, Oslo, 1974), p. 9.

22. *An Energy Policy for Canada*, pp. 189 ff.

23. Krapels, *Controlling Oil*, pp. 20 ff.

24. Evensen, *Instilling om organisasjon for statlige kontinentalsokkelsaker*, p. 29.

25. Thomas Chr. Wyller, 'Oljen og vårt politiske system' in Kari Brunn Wyller and Thomas Chr. Wyller (eds.), *Norsk Oljepolitikk* (Gyldendal, Oslo, 1974), pp. 153–74.

26. James G. March and Johan P. Olsen, 'Organizational Choice under Ambiguity' in Olsen and March, *Ambiguity and Choice in Organizations*, pp. 10–23.

27. Kenneth W. Dam, *Oil Resources* (University of Chicago Press, London, 1976), p. 140.

28. Ibid., p. 130.

29. *Sur les Sociétés Pétrolières*, pp. 224 ff.

30. Dam, *Oil Resources*, p. 113.

31. Ibid.

32. Ibid.

33. *Operations on the Norwegian Continental Shelf*, Report No. 30 to the Norwegian Storting (1973–4), (Ministry of Industry, Oslo, 1974), p. 39.

34. Dam, *Oil Resources*.

35. Krapels, *Controlling Oil*, p. 27.

36. *A Study of UK Nationalised Industries* (National Economic Development Office, London, 1976), pp. 22 ff.

37. 'Norway, a Survey', *The Economist*, 15 November 1975, p. 19.

38. *A study of UK Nationalised Industries*, pp. 22 ff.

39. Ibid., p. 29.

40. Krapels, *Controlling Oil*, p. 35.

41. *Den norske stats oljeselskap*, St.meld.nr. 33 (1977–8), (Ministry of Industry, Oslo, 1978), p. 25.

42. Ibid.

43. *Oil and Gas Journal*, Seventy-Fifth Anniversary Issue (August 1977), p. 491.

44. *Petroleum Industry in Norwegian Society*, p. 9.

45. Eilif Trondsen, 'Søkelys på konsesjonssystemet for oljeblokker', *Sosialøkonomen*, no. 3 (1979), pp. 4–6.

46. *Petroleumsundersøkelser nord for 62° N*, St.meld.nr. 57 (1978–9), (Ministry of Oil and Energy, Oslo, 1979), pp. 81 f.

47. Ibid., p. 82.

48. Trondsen, 'Søkelys på konsesjonssystemet for oljeblokker', p. 5.

49. *Petroleum Industry in Norwegian Society*, p. 14.

6 TAXATION POLICY

Taxation and the Distribution of the Surplus

Aside from depletion policy and licensing, taxation policy has probably given the UK and Norwegian governments the most trouble in their overall oil policies. The taxation of the oil companies operating in the North Sea touches upon a number of intricate legal and economic questions. The governments are thus confronted with a complex set of issues, and must take into account several considerations that are sometimes contradictory. In addition, the taxation of the oil companies is eminently political because of its effects on the bargaining relationship between governments and companies.

Taxation policy can be studied from two different angles. It can be treated as a technical problem involving the development of a rational and equitable system of taxation in order to encourage private investors to follow government policies. Or it can be treated as a political problem, concerning the distribution of the oil rent, the economic surplus given by the difference between the commercial value of the oil extracted and the various costs, where the point is that taxation is a tool for the government-landowner to assert its power, while it is also a political expression of the relationship of strength between the government and private interests.

In this way taxation policy can serve a double purpose, as a tool of economic rationality and as an expression of political power. These two functions do not have to be contradictory, but they operate at different levels, implying quite different methods of reasoning, with opposite approaches to the problem. The technical approach sees taxation as part of a relatively static framework within which private businesses operate. This approach at least implicitly recognises private profits as the driving force behind the development. Consequently, improved performance, for example through a reduction of costs, should be stimulated by and lead to improved private profits. The problem is to adjust profits between the different kinds of fields and operations in order to influence the flows of private capital according to government intentions and guidelines. The political approach sees taxation as a dynamic tool in the hands of the government which can regulate the relationship between itself and private businesses. It also embodies at least an

159

implicit recognition that private profits are created by the extraction of government property, and that improved performance, through declining costs or rising prices, should not first of all benefit private capital. The problem is to have a clear idea of the distribution of the economic surplus, of what private profits are, and how they can be controlled.

These two approaches are not mutually exclusive. The technical approach must also have a certain political evaluation of the level of fair taxation, and consequently at least implicitly deal with the distribution of the rent. The political approach cannot do without technical considerations, i.e. how to implement in practice the level of taxation aimed at giving the desirable distribution of rent. The technical approach is the traditional one in taxing private businesses in capitalist industrial countries. The political approach is much more recent, and in the UK and Norway it has been stimulated by the new situation, with the governments as landowners. The political approach does not always have to be stimulated by considerations concerning the distribution of the rent. Considerations of the impact of the flows of money created by the oil operations concerning the macro-economic interests of the state can also encourage a political approach to the problem. The political approach to oil taxation, therefore, is not necessarily due to ideological considerations, but can be the result of quite practical reasoning. This practical reasoning can, however, lead to a more philosophical or ideological reasoning as well.

Macro-economic considerations have been quite important in shaping UK and Norwegian government thinking on oil taxation. In the UK, it was already clear before the oil crisis that the technical approach to oil taxation might produce quite undesirable effects on the balance of payments. In Norway, the approach to oil taxation appears to have been fairly political from the outset, as concerns for the harmful effects of oil production on certain sectors and regions were explicitly taken into account in the 1960s. Current UK and Norwegian government thinking on oil taxation is predominantly political. However, the UK and Norwegian oil taxation systems consist of a mixture of measures that date from different periods and reflect both the technical and the political approaches to the problem.

Government Considerations

Two periods can be distinguished in UK and Norwegian oil taxation policy; before and after 1973. There is not only a change in tax laws and levels, there is a total shift to a predominantly political attitude in the wake of the oil crisis. This change reflects a discontinuity in the relationship between governments and oil companies in the North Sea. To a large extent the UK and Norwegian governments benefited from the revolutionary action of others. The UK government actually benefited from a revolution that it opposed.

Before 1973 the price of oil was low and there was considerable uncertainty about the reserves of North Sea oil. During this period the major concern of the two governments was to attract foreign oil companies. This concern was stronger and more one-sided in the case of the UK. In addition, the UK government voluntarily observed a high degree of prudence in oil taxation in order not to provide ammunition for the OPEC countries in their demands for higher revenues.[1] While in the UK oil taxation was approached from a strictly technical point of view, in Norway the question of income distribution between the government and the private oil companies was aired in negotiations in 1965.[2]

In the UK, the political approach to oil taxation was forced on the government in 1972–3 by the Parliamentary Committee on Public Accounts.[3] The Committee pointed out that the existing tax rates were smaller than those in almost every other oil producing country in the world. As a result, the private companies in UK waters would have a level of profits unmatched anywhere in the world.[4] In addition, the repatriation of profits by foreign oil companies could lead to an excessive strain on the balance of payments. For these reasons the UK began to take a political approach to oil taxes.

After 1973, the UK and Norwegian governments found themselves in a different world.[5] Attracting private companies to the North Sea was no longer a major concern. The main concern was how to get a fair share of the windfall profits from the OPEC revolution. Thus, the distribution of the oil rent became a predominant issue, and this implied a political approach to oil taxation.

In the UK, even in 1973 the Conservative government was considering a complete change in oil taxation, increasing the government's part of the income.[6] This could have been motivated by practical, macroeconomic considerations, and less by directly political thinking in relation to the oil economy. With the advent of the Labour government in 1974, the British approach to oil taxation seems to have become very similar to the Norwegian one, i.e. to capture a large part of the oil rent.[7] The new tax regimes have been elaborated with close consultations between the two governments, and their effects are largely identical, even if details differ.

The present UK and Norwegian government thinking on oil taxation is focusing two dimensions of the problem: the desirable distribution of income and the desired rate of return on private capital invested. In the wake of the 1973–4 oil crisis, explicit targets were set on both points, but as critical parameters change continuously, a dynamic approach to the issue is needed. This may indeed imply a continuous adjustment of tax rates and allowances. In 1974–5 both governments seem to have agreed that the desirable distribution of income would be in the order of 30 per cent to the companies and 70 per cent to the governments.[8] This was supposed to give a desired rate of return on private capital invested in the order of 19–20 per cent.[9] This, of course, is a static approach to the problem, not taking into account the potential for future price rises, cost escalations or cost reductions. The given distribution of income and the given target for return on private capital coincide under specific conditions. If conditions change, a choice has to be made whether to change aspects of the taxation system and whether to give priority to the target for income distribution or to the target for return on private capital. For example, at higher prices the target for return on private capital may be reached with a different distribution of income. In a recent move to increase its revenues from North Sea oil, the UK government has stated that the desirable distribution of income should give a 75 per cent share to the government and a 25 per cent share to the companies.[10] Implicitly, the rate of return on private capital would remain more or less constant.

The new thinking on oil taxation in the UK and Norway represents a full turn away from conventional taxation philosophy in capitalist industrial countries by considering explicitly the distribution of rent and the rate of return desirable on private capital. Instead of being part of a more or less static framework within which private businesses operate, taxation has become a dynamic tool in the hands of active governments, which can be adjusted according to changing circumstances in order to impose given income and return targets on private capital and in order to implement a given distribution of income between governments and private businesses. This shows how the political approach to oil taxation has become predominant in both countries, and how taxation practically functions to assert the power of governments in relation to private industry. This ability of the governments to control the level of private profits in a capital-intensive industry is quite significant of the changing relationship of strength

between governments and private interests. The change in government
thinking is striking in the case of the UK, where the oil taxation issue as
late as the early 1970s was the victim of imperialist considerations
concerning the defence of private UK oil interests in relation to other
oil producing countries. This led to a confusion of the issue, and it was
perhaps the large-scale nationalisation of oil production in OPEC
countries, reducing the potential for private UK oil companies to
engage themselves directly in production abroad, which helped the UK
government thinking out of its dualism in relation to oil taxation, and
enabled it to see its interests more clearly as those of an oil producing
country.

The Norwegian government was never bothered with the defence of
private oil interests abroad, and could from the outset identify its
interests as those of an oil producer, and thus more easily opt for the
political approach to the problem. Even if the early systems of oil
taxation were never really tried out in practice, as oil production on a
large scale commenced only after 1973, there seems to have been a
striking difference between UK and Norwegian taxation levels. This
could have made the situation more difficult for the Norwegian
government, and in particular could have made the prospects of tax
increases much more delicate. The change in UK oil taxation policy,
after 1974, has obviously also served the interests of the Norwegian
government quite well.

The predominance of political thinking in relation to oil taxation
does not mean that taxation is not also thought of as a technical and
practical problem. In both countries there are several details in the
system of oil taxation which appear not yet to have found a satisfactory
solution. For example, both governments appear to be concerned about
the attraction of small fields and about the effects of cost escalation on
public revenue from oil. But these problems can be seen more as
technicalities within a larger context.

The Early System of Taxation

When the UK and Norwegian systems of oil taxation were first designed,
oil taxation in other producing countries was based upon a complicated
system that involved area fees, bonuses, profit sharing and the use of a
posted price as a tax reference price. This complicated system reflects
the history of the relationship between the producing countries and the
international oil industry in which the producers were gradually able to

impose new levels. Such a complicated system was neither rational nor attractive for a new producing country and the use of many of these tax instruments by the UK and Norway indicates a certain degree of uncertainty and confusion. The newcomers lacked confidence in their ability to redesign a tax system used universally.

The royalty from production is the oldest form of compensation to the landowner and is used primarily with conventional concessions and licences.[11] The usual rate of the royalty had been 12.5 per cent in the United States, in Iran and in most other oil producing countries. In the 1960s some oil producers managed to obtain a higher rate, for example Libya in an agreement with France in 1968.[12]

In the early 1960s the UK set a royalty rate of 12.5 per cent and in 1965 Norway chose a rate of 10 per cent for oil and 12.5 per cent for gas. In 1972 the Norwegian system was changed. In order to make smaller fields more economical the royalty rate for gas was fixed at 12.5 per cent and for oil a sliding scale, varying from 8 to 16 per cent, depending on the volume of the field's production, was chosen.

Area fees are in use in oil producing countries throughout the world. In some cases they provide governments with considerable income and they often increase over time in order to hasten exploration and production and discourage companies from holding large areas for future use. In the UK in the early 1960s area fees averaged about £6,250 per 100 square mile block during the first six years. In the seventh year the fee rises to £10,000 and increases by £6,250 a year until a ceiling of £72,500 a year is reached. In Norway area fees were established in 1965 and set at Nkr.500 per square kilometre for the first six years. After the seventh year they rise by Nkr.500 per year until a ceiling of Nkr.5,000 is reached.

Production bonuses have been rarely used in the North Sea. UK legislation does not even provide for bonuses. In Norway, bonuses have been imposed on a few recent licences, particularly in the blocks where Statfjord has been found.

Profit sharing was first used extensively after 1948 in Venezuela and later in the Middle East. The principle of profit sharing is contrary to Western European taxation traditions, where it is the custom for companies to pay 50 per cent or more of their net income in income taxes. In the UK the notion of profit sharing has never been relevant. In Norway, a form of profit sharing was introduced with government participation on a net profit basis in a number of licences issued in the second round in 1969–70.

State participation on a carried-interest or a joint-venture basis was progressively introduced in the 1950s and the 1960s in a number of oil producing countries. In the UK it was limited to direct licensee participation through the National Coal Board and later through the Gas Council during the mid-1960s. In Norway state participation on a carried-interest basis was first introduced at the second licensing round in 1969–70. It has since become a general feature of Norwegian licensing.

During the 1960s in the UK and Norway there was a government consensus that oil companies should be subject to corporate income taxes. In both countries the rate of company taxation was approximately 45–50 per cent of net income. With the oil industry the situation was complicated by the fact that the big oil companies are vertically integrated so the buyer and the seller usually belong to the same corporation.[13] Thus the transfer price does not necessarily correspond to a market price for oil and it can be manipulated in order to regulate the amount of taxes paid. The situation was further complicated by the fact that in the 1960s most of the world's oil was produced at much lower cost than oil from the North Sea. It came from countries where taxation was not particularly strict.

In Norway in 1965 Parliament passed a special tax law for the offshore oil industry. The law required that oil activities on the continental shelf be subject to the same taxes as corporate activities onshore.[14] There were, however, special provisions that in effect allowed a lower rate of taxation.[15] In 1972 the entire corporate tax system in Norway was altered, and the special provisions for the offshore oil industry were abolished.

In order to avoid oil companies accounting artificially low transfer prices for purposes of tax evasion, and in order to avoid lengthy argument, many oil producing countries had introduced a system of 'posted prices' or administratively fixed prices. The system of posted prices had its breakthrough with OPEC in the early 1960s, and the posted price was used by the OPEC countries to calculate royalties and income taxes. Throughout the 1960s the OPEC posted price was higher than the market price for oil, so the system did ensure a certain minimum of revenues for the OPEC countries.

The pricing issue had an obvious relevance for the UK and Norway. The UK government prior to 1974 never had the legal opportunity to decide the price to be different from the transfer price given by the companies. Norwegian legislation at this time provided for negotiations

between the government and the companies to decide the market price
of the oil produced and consequently its tax base. If the government
and the companies were unable to agree, the government had the right
to establish the value of the oil in accordance with the equitable
market price.[16] It has been argued that the Norwegian government
legally would have had the right to use, for example, the OPEC posted
price as its tax base.[17] It should be underlined that the negotiations
between the Norwegian government and the private oil companies also
concerned the substance of taxation, i.e. income distribution, and that
consequently the Norwegian government reserved to itself the right to
change conditions if necessary.

The problem of extensive loop-holes in the oil taxation system was
recognised in the UK well before the oil crisis, but prior to 1974 the
UK government never had the legal opportunity to decide on the price
base of taxation. Norwegian legislation at this time provided for
negotiations between the government and the companies to decide the
market value of the oil produced, and thus its tax base. In case of
disagreement, the government had the right to decide the value of the
oil itself, in accordance with an equitable market price.[18] It has been
argued that the Norwegian government legally could have used, for
example, the OPEC posted price as its tax base.[19]

Table 6.1: The Early System of Oil Taxation

Form of tax	United Kingdom	Norway
Royalty	12.5 per cent	10 per cent until 1972
Bonuses	None	In one case, 1974
Profit sharing	None	In partial use[a]
State participation on a carried interest basis	None	In partial use[a]
Tax base	Companies' transfer price	Price agreed upon, otherwise set unilaterally by the government
Total government 'take'	50–60 per cent intentionally, but doubtful	Supposedly between 55 and 65 per cent, possibly closer to 40–45 per cent in practice

Note: [a]The government in some licences had reserved for itself the option of
profit sharing and/or state participation.

The Present System of Taxation

The new world oil situation that emerged in 1974 led the UK and Norwegian governments to seek a substantially larger share of the oil rent. They had a choice between designing a new system of oil taxation or adding new features to the old regime in order to get the desired results. The latter method was chosen mainly because it was the easiest.

In the case of the UK the problem was to remedy some structural weaknesses in the existing system as well as to improve existing tax instruments. The UK's Committee on Public Accounts had pointed out that UK corporate tax law had loop-holes when it was applied to oil companies. Generally, UK companies can deduct losses and expenses from their taxable income. Oil companies operating in UK waters could thus deduct losses and expenses accumulated world-wide, including taxes and royalties paid elsewhere.[20] The total taxable income of the oil majors in the UK between 1965 and 1973 amounted to only £500,000.[21] It was feared that the international oil companies might continue to accumulate expenses abroad in order to avoid paying income taxes in the UK.[22] In fact the Committee on Public Accounts considered the general income tax to be of such dubious value in relation to the international oil industry that it proposed a quantity tax.[23]

In order to make taxation more effective and to increase the government's share of the oil rent, the UK and Norwegian governments had several options:

the nationalisation of the oil industry;
the renegotiation of existing licences or the imposition of a desired level of state participation;
the establishment of a government monopoly for the purchase of North Sea oil;
a sales tax on North Sea oil;
a production tax on North Sea oil;
an increased income tax.

Nationalisation is obviously the most secure way to maximise the government's share of the oil rent. It would, however, certainly have alienated the international oil industry and created substantial diplomatic problems with other Western nations, especially the United States. The renegotiation of licences is an uncommon practice in North-West Europe, and the government bargaining position is therefore weak, unless it wants to impose new terms, in which case the same problems arise as with nationalisation. Establishing a government monopoly for the purchase of North Sea oil amounts to a *de facto* nationalisation, and would be considered particularly harmful by the companies because a large amount of North Sea oil is destined for export.

Although taxes give the governments less direct control of the oil produced and are less effective in maximising the governments' share of the rent, they are nevertheless politically more acceptable. A sales tax and a production tax are fairly easy to administer, but they tend to harm the economic viability of marginal fields and tend to be ineffective in capturing the oil rent from large fields. An increased income tax requires more complicated calculations and a more extensive system of administration, but it can be more easily adjusted to fit the government's policy intentions. In the end, both the UK and the Norwegian governments favoured increased income taxes as the solution to the problem.

After extensive deliberations the Norwegian government decided not to impose state participation in existing licences or renegotiate the terms of these licences. Instead it created an additional income tax of 40 per cent to be levied on top of the corporate income tax of 50.8 per cent.[24] Oil company protests and threats to withdraw from Norwegian waters led the Norwegian government to withdraw the proposal. However, a new excess profits tax — the special tax — was introduced on the residual profit after income taxes and royalties. The rate of the special tax was to be set by the government every year, and for 1975 it was set at 25 per cent. Given the repeal of the 40 per cent additional income tax and the renewed negotiations with the companies, the 25 per cent rate will probably last until circumstances change significantly.

At the same time depreciation and investment allowances were modified in favour of the companies. As a result, capital investment in production and transportation may be written off linearly over 6 years, and there is a deduction of 10 per cent from the special tax for capital investment over the past 15 years. For taxation purposes the

Norwegian continental shelf is seen as one unit, so losses from elsewhere cannot be deducted from North Sea income. Exploration costs can be deducted from income from fields already in operation, but as a rule investment can only be deducted from the income that is derived from the field concerned. In addition, 50 per cent of the losses incurred on land or on the shelf in activities other than oil production or pipeline transportation may be deducted from income when computing ordinary income tax. Losses incurred on land may be deducted from future incomes over a period of 15 years.

Norway's introduction of the special tax was accompanied by a new pricing system, in which the government sets the price of the oil. This is called the 'norm price'. It is defined as the real market price of the same type of crude over a given period as determined by independent traders on a free market. It has been explicitly stated that the purpose of the norm price is not to increase taxes but rather to avoid long arguments with the companies.[25] In this respect the Norwegian norm price differs from the posted price used in OPEC countries prior to 1974.

As noted above, the UK government faced the challenge of plugging the loop-holes in the earlier taxation system as well as securing a larger share of the profits. The 1975 Oil Taxation Act was a tax frontier isolating oil company activities in the UK from activities by the same companies that take place elsewhere. This reform satisfied one of the main objections of the Committee on Public Accounts.[26]

At first, in November 1974, the UK government proposed a new Petroleum Revenue Tax that was to be imposed as a flat taxation rate for each field.[27] The new tax was to be deductible from the corporate income tax. The companies argued that such a tax would have a negative effect on marginal fields, and would discourage investment.[28] After extensive deliberations, the proposed tax was changed to be less adverse for marginal fields, and to give better guarantees for invested capital. However, the basic principles of a flat rate and the field being the tax unit were retained. Provisions for transferring losses between fields were liberalised and, as with the previous system, a negative profit could be deducted from income in the next year.

This new Petroleum Revenue Tax was set at a flat rate of 45 per cent of income, net of royalties and operating expenses, but not of interest on loans. To guarantee a minimum return on invested capital the Petroleum Revenue Tax is not imposed until the original capital expenditure plus an extra allowance, a 'capital uplift' of 75 per cent, is earned. The justification for this provision is that interest on loans is not deductible, and so the 75 per cent capital uplift compensates for

this.[29] This provision increases a company's cash flow in the early years
of development, which is often decisive to the economics of a field, as
we have seen earlier.[30] In addition, the Petroleum Revenue Tax will not
be imposed on a field if it reduces the return on capital investment to
less than 30 per cent before corporate income tax as calculated on the
basis of historical capital costs. There is also a ceiling; the Petroleum
Revenue Tax will not exceed 80 per cent of the residual profit accrued
in addition to the initial 30 per cent of historical capital costs.[31]
Finally, each field is granted a deduction from the income taxable by
the PRT to an amount corresponding to the market value of 1 million
tonnes of production a year for each field, limited to a cumulative
total of 10 million tonnes per field as a deduction from the income
taxable by the Petroleum Revenue Tax. The Petroleum Revenue Tax
can also be deducted from the corporate income tax, implying that its
real rate is about half of its normal value. As a safeguard for marginal
fields, royalties may in some cases be refunded by the government,
and this refund is not taxable.

Unlike the Norwegians, the British did not introduce an administratively set price. The oil is to be taxed at market law, and the law
specifies in a detailed way the conditions for an open market price at
an arm's length transaction. The contract price is the only price
reference; the terms of the sale are not to be influenced by any other
commercial relationship between the buyer and the seller; the seller is
to have no direct or indirect interest in the further disposal of the
oil.[32]

In the summer of 1978 the UK Labour government decided to
increase the level of taxation on oil companies.[33] The measure proposed
included a rise in the Petroleum Revenue Tax from 45 to 60 per cent,
a reduction in the production allowance from 1 million to 500,000
tonnes a year, and a reduction in the capital uplift allowance in the
Petroleum Revenue Tax from 75 to 25 per cent. The justification was
based on the government's new goal of a 75 per cent share in North
Sea oil profits. In 1979 the new Conservative government decided to
implement most of the tax changes proposed by Labour, raising the
Petroleum Revenue Tax to 60 per cent, and reducing the capital uplift
allowance from 75 to 35 per cent.

During 1979–80 oil prices in the international market practically
doubled because of the cutbacks in Iranian supplies. Consequently, the
basis of petroleum taxation in the UK and Norway was changed once
more. This time both countries decided to keep the taxation structures
introduced in 1975, but modified some of the rates. In the UK the

Table 6.2: The Present System of Petroleum Taxation

Form of Tax	United Kingdom	Norway
Royalty	12.5 per cent	Oil: 8—16 per cent Gas: 12.5 per cent
State participation	General	General
Taxation unit	Field	Continental shelf
Taxation fence	Continental shelf	Continental shelf
Corporate income tax — deductions	52 per cent Royalties Interest payments Operating costs Capital costs	50 per cent Royalties Interest payments Operating costs Capital costs over 6 years
Loss deductions	No foreign losses	No foreign losses; losses in other activities in Norway at 50 per cent
	Loss deductible following year	Losses deductible following year
Depreciation	Over one year	Over 6 years
Tax base	Market value	Norm price
Additional tax — rate — deductions	Petroleum Revenue Tax 70 per cent Operating costs Royalties 35 per cent of capital costs Value of 0.25 mill. tonnes of oil per field a year, with a limit of 5 mill. tonnes	Special tax 35 per cent Operating costs Royalties, interest payments Yearly $6\frac{2}{3}$ per cent of capital costs over 15 years (100 per cent of capital costs)
Capital tax	None	0.7 per cent of invested capital
Taxes due	Four months after chargeable period	Six months after chargeable period
Advance payment	The greater of 15 per cent of assessed PRT liability or 15 per cent of payment on account for previous chargeable period	None

Petroleum Revenue Tax was raised to 70 per cent in the spring of 1980, while in Norway, it was decided to increase the Special Tax from 25 to 35 per cent. In the UK advance payment of part of the amount due under the Petroleum Revenue Tax was introduced. In Norway, the grace period for payment was reduced from 12 to 6 months after the lapse of the chargeable period. The tax rises in the UK and Norway were considered justified and necessary, by a Conservative and a Labour government, by virtue of the large windfall profits on North Sea oil caused by the price increases of 1979—80. This rapid change in circumstance illustrates governments' problems in keeping up with developments in the international scene which alter the basis of operations of the private oil companies in the North Sea.

Critical Issues in Petroleum Taxation

The 1975 petroleum taxation regimes of the UK and Norway were established to cope with a specific historical situation — the sudden windfall profits from North Sea operations provided by the oil price increase in 1973—4. However, North Sea taxation operates in a dynamic context. Fields are different, some are small, others are large, some are capital intensive, others are less so. Costs develop, but differently from field to field, according to reservoir characteristics, pattern of development, transportation solution and project management. Prices change, nominally as well as in real terms. Companies do not finance development in the same way; some rely upon equity financing while others rely upon external borrowing. The purpose of a taxation regime is partly political and fiscal, to capture a share of the rent for the state; it is partly administrative, to establish a set of conditions influencing private micro-economic decisions in a way that is rational from a social and macro-economic point of view. A key problem in designing a taxation regime for oil companies operating within the concessionary system is that it has to combine the fiscal function and the administrative function. This implies a risk that neither purpose is very well served.

Experience during the period 1975—9 gives a basis for judging the effectiveness of the UK and Norwegian taxation regimes. The critical issues are the following:

income distribution between the state and the companies;
sensitivity to changes in prices and costs;
sensitivity to changes in company financing;
treatment of marginal fields;
flexibility.

At the outset, in 1974–5, the explicit goal of both UK and Norwegian petroleum taxation was to secure a total government take of 70 per cent.[34] Experience indicates that the state's share of income has fallen short of this goal. It has probably varied between 61 and 67 per cent in the UK.[35] In Norway, total government take has been estimated to vary between 57 and 66 per cent.[36] This indicates that governments are less able to reach taxation goals than to define ambitions. It also indicates that private industry may be better than governments in discovering loop-holes and special provisions in taxation systems. Finally, it may indicate that the basis for tax calculation has changed. In any case, the performance of the new tax systems has been less than satisfactory from the government point of view.

Neither system is very sensitive to changes in the price of oil (for practical purposes, in the following the Norwegian system will be considered with the maximum allowable dividends declared).[37] In the UK system the progressive effect of the Petroleum Revenue Tax is largely neutralised by other provisions.[38] In the Norwegian system, under the set of rules applied between 1975 and 1979, marginal tax rates were slightly higher than average tax rates on small and large fields, indicating some degree of progressive income taxation, but on medium-sized fields the opposite was the case, indicating some degree of regressive taxation.[39] Consequently, rising oil prices have resulted in a higher government take on small and large fields, but a lower take on medium-sized ones.

The effects of escalating costs seem to have been less uniform. In the UK system, cost escalation seems to have led to a slight reduction of the private share, and a slight increase in the government take, so long as the annual limit of the Petroleum Revenue Tax had not been reached. Once this limit had been reached, the position was reversed. Thus, up to a point, cost increases have a negative impact upon the relative share of private profits. Once this point is reached, this negative impact is reversed. In the Norwegian system, the effects of cost escalation seem to be equally complex, with a general tendency for cost escalations to reduce the government take rather than private income.[40] The main reason is that capital costs can be deducted again from the tax base of the Special Tax, after corporate income tax has been calculated. Thus, with the Norwegian system, cost escalations to some extent correspond to an investment in reducing the tax base, and it is doubtful whether cost escalations on the average have a net negative impact upon private income, at least under the set of rules valid until 1980, when the capital cost deduction from the Special Tax base was reduced from 150 to 100 per cent.

The effects of a combination of cost escalation and price rises have not yet been fully analysed, but this was the situation in the North Sea in the late 1970s. Thus, some tentative conclusions can be suggested. In the UK system, this combination of factors seems to lead to a slight redistribution of total income in favour of the government. However, compared to the initial outlook, before costs and prices started to rise, the net private income and return on private capital invested has increased considerably. In the Norwegian system, this set of circumstances seems to produce more complex effects. On smaller fields and medium-sized fields, there may be a relative redistribution of income in favour of the companies. On larger fields, this effect may be more varied. In both cases private income and return upon private capital invested have increased considerably compared with the initial outlook. Generally, in the Norwegian system, the high capital cost deduction from the Special Tax base means that the combination of rising costs and rising prices benefits the private investor.

In both systems the government take is reduced by increased external borrowing by companies because interest payments may be deducted from taxable income. This effect is generally stronger on larger fields than on smaller ones.[41] Also, it is more visible in the Norwegian system where finance charges may be written off before calculating the base of the Special Tax, whereas this is not the case with the UK Petroleum Revenue Tax. In fact, the larger the field, the more visible this difference.

Both systems have been criticised for not devoting sufficient attention to marginal fields.[42] Under rules valid from 1975 to 1980, the total government take in both the Norwegian and the UK systems tended to increase with declining field sizes.[43] This implied an irrational regressive taxation from the point of view of resource management. With North Sea development advancing and with the larger fields in the southern waters being depleted, the question of the economic attraction of marginal fields in southern waters as well as cost intensive fields in northern waters becomes more acute. In the UK system marginal fields can be made more attractive through the discretionary waiving of government royalties. In the Norwegian system similar provisions have been introduced. Still, the problem of marginal fields remains unsettled. An important feature in this respect is the fact that the Petroleum Revenue Tax and the Special Tax are relatively more important for the taxation of smaller fields than for larger fields.[44]

This raises the question of the flexibility of the two taxation systems.

In the UK system changes in the Petroleum Revenue Tax do not readily produce changes in the distribution of income between government and companies. Whereas in Norway changes in the Special Tax can produce a marked difference in the distribution of income. In this way, the Norwegian system of petroleum taxation appears more flexible. On the other hand, the British system produces less variations in the distribution of income between different fields and is probably less sensitive to changes in external factors, and in this way appears more stable. This contrast between flexibility and stability leads to the structural differences between the two taxation systems.

The Problem of Taxation Structure

As we have seen, the historical point of departure for special petroleum taxation in the UK and Norway was the oil price rise of 1973—4. Prior to this, the petroleum sector had been subject to the same corporate income taxation as other businesses in the two countries, even if in Norway the government had declared some explicit tax intentions concerning the petroleum sector. In the aftermath of the 1973—4 oil price increase, corporate income tax was no longer considered an adequate instrument for taxing the domestic petroleum sector, and for securing for the government a good share of the additional rent given to North Sea oil by the OPEC price increase. The problem was to secure part of the oil companies' residual income after paying corporate income tax to the government. In this respect two choices were apparent:

to tax the residual income after corporate income tax through an excess profits tax;
to reduce the income base upon which corporate income tax was levied through a special excise tax.

The Norwegian and the UK governments each selected one solution. In Norway, it was decided to apply a new tax, the Special Tax, upon the residual income of oil companies after the corporate income tax had been applied. Thus, the Special Tax in the Norwegian system is essentially an excess profits tax.[45] In the UK, it was decided to reduce the income base upon which the corporate income tax was applied, through a special levy. Thus, the UK Petroleum Revenue Tax is essentially an excise tax. In this way, the structures of the two taxation systems differ profoundly. In both cases, the petroleum taxation system was modified in a number of ways in order to improve flexibility and protect private profits within certain limits.

As already mentioned, both the UK and the Norwegian petroleum taxation systems appear to have fallen short of the targeted government take announced in 1974–5. In the spring of 1980, the intention of both governments appears to be a government take in the neighbourhood of 80 per cent, or even a few percentage points more. This may seem excessive compared to the target of 70 per cent announced five years previously, but a company take of 20 per cent at 1980 prices may well correspond to, if not excede in real terms, 30 per cent at 1975 prices. Furthermore, experience so far suggests that both governments will have difficulty in actually achieving the intended tax level of 80 per cent.

With hindsight, the 1975 tax regimes of the UK and Norway appear as defensive measures. They do not appear as aggressive policies aimed at exorbitant government takes, as they were sometimes presented by the oil companies, and by the governments themselves. At least in Norway, the total government take under the 1975 system did not appear to be much higher than projected under the old system. The intention prior to 1973–4 was a total government take in the order of 55–65 per cent.[46] In fact, at least until 1980, 57–66 per cent was apparently what was achieved.[47] In the UK there has been greater discontinuity of tax policies and tax systems, but compared to pre-1973 Norwegian intentions, the post-1975 UK government take is not extraordinary.

The remarkable fact is that the new oil taxation regimes of both countries represent important innovations in taxation policy. This was absolutely necessary, at least in Norway, so as not to change drastically the distribution of income, but rather to keep the *status quo*. This also demonstrates that the concessionary system presents some awkward problems of control and that it is difficult to combine the concessionary system with a taxation system that responds to fiscal and political concerns as well as technical and administrative issues.

Similarly, the modifications in tax rates introduced in the UK and Norway in 1979 and 1980 also appear as defensive measures designed to maintain the *status quo*. Given that the oil price increase of 1979–80 will not be the last this century, it is a fair bet that petroleum tax rates in the UK and Norway will change again. But they are likely to change after the prices rise, indicating the defensive and conservative character of the tax modifications. This situation is unsatisfactory for both

governments and companies. Governments have a constant fear that
their take will fall short of declared targets, particularly in situations of
market turbulence, exposing them to domestic criticism. Companies
are forced to operate in a politically uncertain environment, never
really knowing what the tax rates are going to be over the life of a
field.

Indeed, neither system seems able to cope with changes in key
external factors. In the UK system the annual limits on the Petroleum
Revenue Tax may put an effective ceiling on the total government take
after a new price increase.[48] In the Norwegian system, marginal tax
rates are supposed to be progressive after the modifications of 1980
but in the event of further price rises fresh modifications will be
required.[49] Similarly, in the UK system, the annual limits on the
Petroleum Revenue Tax may also reduce government take in the
event of significant cost escalation. In Norway the modifications of
1980 have hardly removed the risk that the tax base of the Special
Tax may be seriously eroded by escalating costs. Thus, in both systems,
companies may continue to experiment largely at government expense,
particularly if higher oil prices are anticipated. This illustrates a
fundamental weakness of both systems, but perhaps most acutely in
the Norwegian case.

It can be argued that the economic attraction of marginal fields
might be improved through modifications of the Petroleum Revenue
Tax or the Special Tax. But such modifications would also influence
the distribution of income on larger fields. This illustrates the limita-
tions of a uniform taxation regime. This observation is valid for both
systems. Indeed, within the concessionary system it appears hardly
feasible to design a petroleum taxation regime that is rational and
equitable for different fields with different characteristics and costs.
The Norwegian petroleum taxation proposal of 1980 explicitly states
that it is hardly feasible to combine a high marginal tax on petroleum,
with a small share of incremental capital costs being borne by the
government.[50] The point is that a progressive taxation on petroleum
although desirable from the point of view of a higher government take
on windfall profits and in stimulating the development of marginal
fields, can be counterproductive in stimulating cost consciousness.
These observations apply especially to the Norwegian system of taxing
the residual income after corporate income tax, as this residual tends
to be highly sensitive to changes in costs, prices, field sizes, etc. They
apply to a lesser extent to the UK system of regulating the income
base upon which corporate tax is applied, but in principle the same

contradictions concerning progressive taxation are valid here as well.

As a general critique it should be added that both the UK and the Norwegian systems are in their current form extraordinarily complex, and perhaps unnecessarily so. Complexity makes it difficult for outside observers, and probably for the civil service and industry as well, to assess correctly how the petroleum taxation functions when important parameters change. This raises the question whether alternative taxation systems are feasible, that possibly also might be simpler.

A production tax on petroleum was mentioned as an alternative to the Norwegian petroleum taxation proposal of 1980, but was discarded because of complexity and undesirable effects upon marginal fields.[51] However, a general production tax on petroleum could be used to shelter the North Sea operations, partly or wholly, from price changes in the international market, with the differential accruing to the government. Such a production tax need not be applied specifically to individual fields, but could be applied to the North Sea, or either sector, and adjusted according to the real price development for oil after an initial year, e.g. 1974. A uniform production tax, with no safeguards, would have an undesirable effect upon the development of marginal fields. A production tax with safeguards, for example in the form of deducting capital costs entirely, would not stimulate cost consciousness. An intermediary solution, with a provision for deducting 50 per cent of capital cost, might be a sensible trade-off. Structurally, such a solution would resemble the British system. However, it ought to be much simpler and easier to govern, as it would do away with the annual limits and royalties, that are a vestige of the excise taxes imposed in desperation by weak governments upon strong foreign oil companies.

Notes

1. *First Report from the Committee on Public Accounts* (HMSO, London, 1973), p. xi.

2. Jens Evensen, *Oversikt over oljepolitikse spørsmål* (Ministry of Industry, Oslo, 1971), p. 63.

3. *First Report*, p. xxxiii.

4. Ibid., p. xxiii.

5. Ot.prp.nr. 26 (1974–5) *Om lov om skattlegging av undersjøiske petroleumsforekomster*, m.v. (Ministry of Finance, Oslo, 1975).

6. D. I. Mackay and G. A. Mackay, *The Political Economy of North Sea Oil* (Martin Robertson, London, 1975), pp. 32 f.

7. Ibid., p. 41.

8. Ot.prp.nr. 26 (1974–5), p. 23.

9. Ibid.

10. *The Economist*, 5–11 August 1978, p. 73.
11. Evensen, *Oversikt*, pp. 34 f.
12. Ibid., p. 40.
13. Ibid., pp. 44 f.
14. Eivind Erichsen, *Hydro-Carbon Taxation Policy in Norway*, speech at the Second Scandinavia and North Sea Conference, Oslo, 1–2 September 1965, p. 2.
15. Ibid.
16. Evensen, *Oversikt*, p. 47.
17. Ibid., pp. 46 f.
18. Evensen, *Oversikt*, p. 47.
19. Ibid.
20. Mackay and Mackay, *The Political Economy of North Sea Oil*, p. 32.
21. Ibid.
22. Ibid.
23. *First Report*, p. iii, and Jon R. Morgan, 'The Problems and Promises of Petroleum Revenue Tax' in *The Taxation of North Sea Oil* (Institute for Fiscal Studies, London, 1976), p. 7.
24. Kenneth W. Dam, *Oil Resources* (University of Chicago Press, Chicago, 1977), p. 69.
25. Erichsen, *Hydro-Carbon Taxation Policy*, p. 3.
26. R. F. Hayllar and R. T. Pleasance, *UK Taxation of Offshore Oil and Gas* (Butterworth, London, 1977), pp. 148 ff.
27. Dam, *Oil Resources*, p. 125.
28. Ibid.
29. Ibid., p. 127.
30. Ibid.
31. Ibid., p. 126.
32. Hayllar and Pleasance, *UK Taxation of Offshore Oil and Gas*, pp. 32 ff.
33. *The Economist*, 5 August 1978, p. 73.
34. Erichsen, *Hydro-Carbon Taxation Policy*, p. 3.
35. Colin Robinson and Jon Morgan, *North Sea Oil in the Future* (Macmillan, London, 1978), pp. 95 ff.
36. Dam, *Oil Resources*, p. 69.
37. Robinson and Morgan, *North Sea Oil*, pp. 95 ff.
38. Morgan, 'The Promise and Problem of Petroleum Tax', p. 28a.
39. Ot.prp.nr. 37 (1979–80) *Om lov om endring i lov av 13. juni 1975 om skattlegging av undersjøiske petroleumsforekomster m.v.*, Oslo, 1980, Ministry of Finance, pp. 122 ff.
40. Robinson and Morgan, *North Sea Oil*, pp. 95 ff.
41. Ibid., p. 102 ff.
42. Dam, *Oil Resources*, p. 102.
43. Robinson and Morgan, *North Sea Oil*, pp. 95 ff.
44. Ibid., p. 96.
45. Ibid., p. 95.
46. Evensen, *Oversikt*, p. 43.
47. Dam, *Oil Resources*, p. 69.
48. Morgan, 'The Promise and Problem of Petroleum Tax', p. 28a.
49. Ot.prp.nr. 37 (1979–80) *Om lov*, pp. 134 ff.
50. Ibid., pp. 112 ff.
51. Ibid., pp. 134 f.

7 THE IMPACT OF ECONOMIC POLICY

North Sea oil has changed the premisses of economic policy in the UK and Norway. This basic structural change has been both positive and negative. Among major advantages are the following.

(1) The transition from being a net importer of oil to being self-sufficient or a net exporter has a positive effect on the balance of trade, permitting a quicker expansion of production and consumption than would otherwise have been possible.

(2) Domestic supplies of oil and natural gas provide the flexibility for long-term energy planning and the establishment of a competitive advantage in energy-intensive industries.

(3) The development of a petroleum industry creates a domestic market for advanced goods and services that are specially related to oil production and are in general capital-intensive with a high content of new technology. This new market in turn creates a basis for an internationally competitive position in this field.

(4) The greater demand for productive factors that accompanies economic expansion causes domestic industry to rationalise production and improve its efficiency. In addition, oil revenues improve the availability of capital, permitting increased industrial investment and greater industrial specialisation, thus adding to the economy's overall international competitive position.

(5) Finally, public revenues from oil can finance tax cuts, increased public spending, improved social services and a more vigorous regional policy.

These advantages are counterbalanced by a number of serious risks.

(1) The oil industry can overheat the economy and fuel inflation, which can cause the country's exchange rate to rise, which can in turn hurt traditional exports and stimulate imports.

(2) Vast domestic supplies of oil and natural gas can lead to energy waste through low domestic prices and a lax attitude towards energy conservation.

181

(3) Insufficient domestic expertise in the new market for support services and technical products needed for the petroleum industry can lead to a tremendous reliance on imports. This also means that domestic industry loses its chance of acquiring new skills.

(4) The increased cost of productive factors can make domestic industry less competitive at home and abroad rather than rationalising and improving productivity. Oil revenues can also be used to shore up ageing industries through subsidies, rather than being used for industrial investment, specialisation and reorganisation that would enhance international competitiveness.

(5) A poorly planned and co-ordinated oil industry can lead to such high social and environmental costs that a large proportion of oil revenues is required to neutralise these negative effects.

Thus, for the UK or Norway to benefit from the potential advantages of the new oil industry, a delicate and careful balancing of economic, energy, industrial, social and regional policies is necessary. In this regard, macro-economic policy is perhaps the most important tool in the hands of the government. However, macro-economic measures must be supplemented by more specific sectoral policies. It is essentially up to the governments whether their new-found oil is turned into an advantage or a liability.

Economic Policy Since the Second World War

During the first phase of North Sea oil development — until the oil crisis — the UK and Norwegian economic contexts were strikingly different. In the period after the Second World War the UK had a sluggish and uneven rate of economic growth, with chronic balance of payments problems, a generally low rate of investment and serious labour problems. During this same period Norway had a high and remarkably stable rate of economic growth, with no serious balance of payments problems, an extremely high rate of industrial investment, and remarkably good labour relations.[1] The relative positions of the two countries have changed significantly. Living standards in the UK in the early 1950s were among the highest in Western Europe,[2] and now they are among the lowest. Norwegian living standards were comparatively modest in the early 1950s, and around 1980 they are among the world's highest.

These differences in performance can be explained by a number of different factors. There is an obvious difference in the rate of investment. The UK rate of investment has increased slowly from about 16 per cent of GNP in 1950 to about 22 per cent in the mid-1960s.[3] The Norwegian rate of investment since 1945 has consistently been between 30 and 35 per cent of GNP.

In the 1960s and early 1970s the UK has been notorious for its high frequency of strikes, while Norway throughout the post-war era has had one of the calmest industrial scenes in the world. Poor industrial relations have a negative effect on productivity. This reduces the return on capital, and tends to deter further investment. This difference in labour relations explains part of the difference in economic performance between the two countries.

The different rates of investment and the differences in labour relations are symptoms of different macro-economic policies. UK macro-economic policy in the 1950s and 1960s was to a large extent oriented towards international goals, such as the maintenance of the UK's position as a world power and the defence of the pound as a reserve currency.[4] Norwegian macro-economic policy, by contrast, has been essentially directed at achieving a high and stable rate of domestic economic growth without aggravating the decline of older industries or problems of internal migration.

These contrasting policy orientations have produced quite different results. In the context of the UK the defence of the pound meant sacrificing domestic economic growth whenever it threatened the position of the pound. This naturally caused sluggish and highly uneven growth, and discouraged investment and modernisation in UK industry. Furthermore, the slow and uneven growth did not produce the new resources required to permit an equalisation of incomes and living standards without reducing real incomes for certain groups. This slow growth produced the labour militancy and the high strike frequency of the late 1960s and early 1970s, which appeared only after a long period of economic stagnation.

In the Norwegian context, high and stable growth was encouraged through investment and the modernisation of industrial management. This economic growth allowed for the equalisation of incomes and living standards without unduly reducing real incomes. This created an atmosphere of social harmony, which stimulated more industrial modernisation and growth.

In the late 1960s both the Labour Party and the Conservative Party realised that the UK could no longer afford to act as a major inter-national financial power.[5] Since the 1967 devaluation of the pound, both parties have given a hesitant priority to domestic growth. However, the damage done to the UK economy was considerable. Its physical plant was to a large extent outdated, causing output to lag behind its principal competitors. The antiquated machinery was combined with old-fashioned management. Labour relations had become perhaps the worst in Western Europe, with class conflict intensifying and reaching a level that seemed to threaten the political basis of UK capitalism. This situation required an economic miracle that could quickly improve the balance of trade, foster a higher rate of economic growth, provide a new market and new incentives for domestic industry, stimulate increased industrial investment, and finally permit an increase in private consumption together with improvements in social services and regional policies.

In the late 1960s, the Norwegian rate of economic growth had picked up, the balance of trade improved, and the movement of labour from agriculture and fishing to other sectors of the economy accelerated. In this situation of prosperity increasing attention was given to balancing economic growth against individual well-being. The productive apparatus of the Norwegian economy was one of the most modern in the world. Management was able to cope with most of the problems of the decentralised economy and industrial relations were remarkably smooth. Norwegian capitalism not only appeared economically sound, but it also seemed to be politically settled. In this situation there was an obvious risk that the introduction of a new capital-intensive industry might harm this successful economy. This in turn could increase the level of political conflict in Norwegian society and seriously weaken the political consensus established in the post-war period.

Macro-economic Policy and Oil

The differences in the post-war economic positions of the UK and Norway are further accentuated by the size of the new oil industries in relation to the traditional economies. The UK has a population about 14 times the size of Norway's. In 1973, UK GNP was about 9 times that of Norway. This results in a difference in GNP *per capita* in 1973 of about 57 per cent in Norway's favour ($2,692 against $4,216). This difference in size means that the same level of production has a very different impact on the two economies. Whereas at 1980 prices the oil and gas sector might perhaps add 10 per cent to the UK gross national product, in Norway it might increase GNP by as much as 25 to 50 per cent.

One of the more remarkable aspects of the history of North Sea oil is that it started to flow in significant quantities just after the oil price rise. This good fortune gave higher returns to companies and governments, making the relationship between oil policy and macro-economic policy more important, and, compared to most other OECD countries, allowing greater freedom of action in external economic policy. At the same time, the UK and Norway, given their open economies and dependence upon foreign trade, have been hit by the inflationary problems and sluggish economic growth of the OECD area since 1973. Consequently, 1973–4 appears as a historical divide for the UK and Norwegian economies. On one hand, the advantage of being an oil producing country was accentuated; on the other hand the non-petroleum sectors of the two economies have fared worse than expected previously; thus, the petroleum sector has become the motor force in the UK and Norwegian economies. This represents a qualitative change for economic policy, with the key task now being to adjust the non-petroleum sectors to a macro-economic environment dominated by the petroleum sector.[6]

A. United Kingdom

Becoming self-sufficient in oil clearly enhances the UK's prospects for economic growth and a stable balance of trade.[7] This should in theory permit a more dynamic macro-economic policy, but this does not seem to have happened. Part of the reason in the mid-1970s may have been the legacy inherited by the Labour government in 1974 from its Conservative predecessor. In 1973 the UK GNP grew by 4.4 per cent but this also provoked a serious deficit in the balance of trade, due partly to high oil imports during the last two months of the year, and partly to economic expansion.[8]

In 1974 and 1975 there was some redistribution of income in the UK economy, with the share of wages and salaries increasing and the share of profits and related income decreasing. Wages increases were even higher than the high prices rises. In late 1976 the UK government was forced to negotiate a loan from the International Monetary Fund. One of the conditions of the loan was that the growth of wages be kept below the growth of prices.[9] These conditions were satisfied in 1977, with real wages declining, and in 1978, economic growth was

Table 7.1: UK Economic Indicators 1973–9

Per cent	1973	1974	1975	1976	1977	1978	1079*
GDP Growth	5.2%	0.2%	−1.6%	3.1%	1.8%	3.4%	0.8%
GDP Growth excluding oil and gas			−1.6%	2.6%	0.8%	2.7%	0.0%
Unemployment	2.7%	2.6%	3.9%	5.3%	5.8%	5.5%	6.0%
Rate of inflation	9.2%	16.0%	24.2%	16.5%	15.9%	8.3%	13.5%
Parts of GDP: Employment income	67.6%	71.4%	73.2%	71.2%	69.3%	69.3%	69.5%
Gross trading profits and surplus	17.3%	16.4%	14.3%	16.1%	16.8%	16.0%	18.1%
Private consumption	62.3%	60.9%	60.2%	59.2%	59.7%	60.2%	60.5%
Public consumption	18.3%	19.4%	21.8%	21.4%	20.7%	20.5%	21.0%
Gross investment total	19.5%	19.8%	19.4%	18.9%	18.4%	18.3%	17.1%
oil	0.3%	0.7%	1.3%	1.7%	1.5%	1.4%	1.6%
other	19.2%	19.1%	18.1%	17.2%	16.9%	16.9%	15.5%
Exports–Imports	−3.3%	−6.3%	−3.1%	−3.2%	−1.6%	−0.9%	−0.8%
Trade balance related to oil	−1.3%	−4.0%	−2.9%	−3.2%	−1.9%	−1.2%	−0.5%

*Preliminary figures, first three quarters, seasonally adjusted.
Sources: *Central Statistical Office Monthly Digest of Statistics*, March 1980, and *OECD Economic Surveys United Kingdom*, February 1980.

able to pick up. By 1978 UK oil production reached 58 million tonnes, corresponding to more than one half of domestic consumption, so that the trade deficit related to oil was reduced considerably. Thus, the basis for a more sustained economic growth, without unduly compromising the external economic situation, seemed reasonably assured for following years.

However, the Conservative government elected in the spring of 1979 opted for a radically different economic policy. As net self-sufficiency in oil was approaching, the trade deficit related to oil was reduced to negligible proportions. Compared to previous years and compared to most other West European countries, this should have given enlarged freedom of action in macro-economic policy. This freedom of manoeuvre was not used to stimulate investment and economic growth, at least not in the first instance. Instead, priority was given to a tight monetary policy and a rising currency rate, stimulating domestic inflation and imports, but with an adverse effect upon exports, investment, economic growth and employment. It is possible that the tight monetary policy will give some of the expected results, i.e.

bringing inflation down, rationalising the industrial structure and laying the basis for renewed investment and economic growth. Otherwise, there is a risk that the tight monetary policy, will have neutralised the potential for additional economic growth presented by the petroleum sector.

At pre-1979 prices the petroleum sector was small in relation to the total UK economy, and it was thought that it would add perhaps 6 to 8 per cent GNP in the 1980s.[10] Because of imports required for oil and gas development and the remittances of foreign firms, the petroleum sector had a negative effect on the current balance during the build-up phase in the 1970s. The yearly net effect was expected to become positive in 1979, with the cumulative effect becoming positive in 1983.[11] At 1980 prices, the petroleum sector could add perhaps 10 per cent to GNP in the 1980s, with the cumulative effect becoming positive perhaps by 1981.

According to earlier calculations, by 1981, the UK production of oil and gas should permit a 20 per cent increase in imports in relation to a constant target for the 1978 balance.[12] At current prices, this figure might exceed 30 per cent, indicating a substantial improvement in the freedom of action in macro-economic policy. Given the traditionally high demand for imports, there is a risk that a modest growth in real wages could use up most of the new import potential. This raises the issue of the competitiveness of UK industry and the need for new industrial investment. Channelling the new financial resources into private profits could be equally risky, given the reluctance of many UK capitalists to invest and the potential for social conflict that might be generated by a more unequal distribution of income. This dilemma makes the need for a balanced public industrial policy more acute. Such a policy could be facilitated by the fact that a large part of petroleum revenues go to the state.

In the first public document discussing the use of oil revenues, the last Labour government gave a high priority to industrial investment and the restoration of the competitive position of UK industry.[13] There was an explicit warning against using oil revenues to finance an immediate rise in living standards and a rapid expansion of public services.[14] The current Conservative government has a preference for using oil revenues to finance general tax cuts, hoping that this will stimulate spending, investment and economic growth outside the petroleum sector. The success of this policy largely depends upon the stimulating effects being kept within the UK economic circuit, and not 'leaking' abroad. This again, can be seen as contrary to the policy of

tight money and a high exchange rate, unless there is an explicit desire
to encourage investment outside the UK economic circuit. In this
perspective the difference between Labour and Conservative policy to
some extent appears as the opposition between industrial capital with
a domestic orientation and finance capital with an international
orientation.

B. Norway

Since the early 1970s, Norwegian macro-economic policy has been
directly affected by the prospect of new sources of income from oil.
Even before the 1973 price rise, the two fields then being developed
were expected to add 6 per cent to GNP by 1977.[15] The Long Term
Programme, published in the spring of 1973, declared that the
petroleum sector would add 1 per cent a year to GNP.[16] At this time
there was a trade surplus amounting to 2 per cent of GNP. Consequently, the Programme, which covered the period until 1977,
projected an economic expansion that gave industrial investment a
higher priority than private or public consumption, and aimed at
maintaining the trade surplus.

A subsequent plan was presented by the Labour government in the
winter of 1974.[17] This was an entirely new plan integrating oil revenues
into the Norwegian economy, and defining general macro-economic
policy in relation to oil. The new plan, based on the new prices of oil
and gas, recommended a two-target policy: a cautious approach to oil
and to the domestic use of oil revenues in order to avoid overheating
the economy, combined with a high rate of economic growth, with
private and public consumption getting a higher priority than in the
Programme presented a year earlier.

This plan is a basic document in recent economic and oil policy. It
set a moderate rate of development for the petroleum sector, limiting
production to 90 million tonnes of oil equivalents a year, a level that
Norway expected to reach shortly after 1980. For petroleum revenues,
it was recommended that only about half of the public income from
petroleum be spent in the domestic economy.

The plan gave particular attention to the transition of labour
between industries and adaptations to new regional and social problems.
The plan was based on detailed calculations and projections about the
position of industries exposed to foreign competition and those
sheltered from it. The labour needs of the oil industry were expected
to cause few problems because of the low employment in that industry.
Through increased public spending a high demand for labour was

expected, which would increase wages. These higher production costs would stimulate industries exposed to foreign competition to rationalise and increase productivity, thus freeing labour for the rapidly expanding industries sheltered from foreign competition that would be growing because of increased domestic demand.

Table 7.2: Norwegian Economic Indicators 1973–9 (per cent)

	1973	1974	1975	1976	1977	1978	1979*
GNP growth rate	4.2%	5.3%	5.5%	6.8%	3.6%	3.5%	3.0%
Unemployment	0.8%	0.7%	1.2%	1.1%	0.9%	1.3%	–
Rate of inflation	7.5%	9.5%	11.7%	9.1%	9.1%	8.1%	4.4%
Wages increase		14.2%	22.0%	8.3%	10.8%	7.9%	4.0%
Parts of GNP:							
Wages costs	76.1%	75.6%	78.0%	79.4%	81.5%	79.2%	75.3%
Profits independents (a)	4.9%	4.7%	4.5%	5.2%	5.8%	5.5%	4.5%
Capital income etc.	13.1%	13.0%	11.4%	9.2%	6.4%	8.1%	10.7%
Private consumption	53.9%	53.2%	54.6%	55.0%	56.1%	53.8%	49.3%
Public consumption	16.2%	16.1%	16.9%	17.4%	18.5%	18.5%	19.8%
Gross investment total	30.4%	33.6%	35.2%	37.1%	35.9%	28.7%	28.3%
Gross investment oil	2.8%	4.3%	5.1%	6.3%	7.2%	4.4%	2.9%
Gross investment rest	27.6%	29.3%	30.1%	30.8%	28.7%	24.3%	28.3%
Exports–Imports	0.5%	–2.9%	–6.7%	–9.5%	–10.5%	–10.6%	2.6%

*1979 figures provisional.
(a) independents in farming, fisheries, etc.
Sources: Norwegian National Budgets 1977 and 1980.

In the plan the pre-1974 rate of economic growth in the non-oil economy was assumed to continue, making the risk of overheating the domestic economy a permanent hazard for Norway. A policy of importing large numbers of foreign workers might have eased the problem, but this was considered to be a politically unacceptable alternative. The prospect was that by 1977 Norway would already have a considerable balance of payments surplus and the major problem would thus be to absorb the increasing oil revenues. The plan antici-pated a phase of intense economic growth up until 1977 because of the rapidly expanding petroleum sector. After 1977 growth rates were expected to level off. Consequently, the risk of overheating the economy was thought to be most serious in the short term.

At the social and political level, the plan had a clear social democratic orientation. The petroleum industry was to be kept under firm public control, with Statoil being the main instrument for implementing the government's oil policy. The new wealth was to be used to make Norway 'a qualitatively better society', and the plan announced a series of social, regional, environmental and industrial policies to achieve this goal. The government also announced its intention to use part of the oil revenues to increase its participation in Norwegian industry, and in particular to buy up several foreign industrial interests.

In the spring of 1975 another report was published on oil and economic policy.[18] It anticipated high growth rates until 1980, and also discussed various long-term options. At this point, Norway's oil policy and economic policy received widespread international attention. A country report of the OECD described the cautious attitude to oil production and the domestic use of oil revenues as signs of economic maturity. It predicted that by adapting its economy to the new conditions created by oil, Norway would go through a 'passionate and stimulating experience'. Norway was said to have husbanded her oil potential with great skill and foresight,[19] and was expected to become the Kuwait of the north within a few years.[20] It was also described as a good example of a successfully run economy.[21]

Norway's main problem in 1974 and 1975 was that, with half of its GNP oriented towards foreign trade, it was highly vulnerable to the international economic recession. The Norwegian government at first thought the recession would be fairly short, and could be successfully encountered by anticyclical measures that would maintain production and employment. Its policy measures included special loans to build up stocks, investment incentives and direct subsidies. This anticyclical policy required deficit spending and foreign borrowing.

In addition, the first phase of the oil development was financed by foreign borrowing, as domestic capital markets were insufficient. In a later phase, development might to a large extent be financed from the government's anticipated financial surplus. In this situation it was seen as appropriate to use part of a future financial surplus in order to overcome short-term problems through increased foreign borrowing. Norway's shipowners were also hard hit by the recession, in particular the Norwegian tanker fleet. A large-scale rescue operation was initiated in order to prevent a large part of the Norwegian fleet from being sold abroad at low prices. This operation was also financed from the anticipated future financial surplus through foreign borrowing. In the 1970s the Norwegian government has been considered extremely creditworthy because of its oil and has never had any problems getting foreign loans. However, as the international recession continued, it became increasingly evident that the chosen policies were inappropriate. The anticyclical policy had negative effects on industrial performance and foreign debt grew to alarming proportions.

As the recession continued, the anticyclical policy developed into a massive subsidisation of industry, employment and private consumption. By the beginning of 1978 about a quarter of the jobs in manufacturing depended on government subsidies. Wage settlements from 1971–3 created a stagnation or decline in real income for large groups of workers despite an increase in profits and related incomes in the same period. Consequently, political pressure for improving real wages grew, explaining the high wage increases in 1974 and 1975. In 1974 Norwegian wage increases were accepted without significant opposition from the employers. In 1975 the government intervened.

During 1974 and 1975 the growth of wage costs averaged 14 per cent more for Norwegian industry than for its principal foreign competitors. Nominal wages increased by 51 per cent from 1974 to 1976. During the same period prices rose by 33 per cent, giving a net increase in real wages of 14 per cent. In 1976 and 1977 the government participated directly in the wage settlement in order to moderate the increase in nominal wages. This was only a partial success. Between 1976 and 1978 wages drift added more to nominal wages than did the formal settlements. From 1976 to 1978 wages increased by 19 per cent, 12 per cent of which was due to wages drift.

In addition to domestic labour cost pressures, the position of Norwegian industry was also affected by exchange rate policy and the continued international recession. In 1973 the Norwegian currency was revalued by 5 per cent in anticipation of a gradual rise in the value of the currency because of oil revenues. Throughout this period the Norwegian currency has been tied to the German mark through the European currency snake. Because of the differences between German and Norwegian economic policy prior to 1977, there was in effect an increase in the value of the Norwegian kroner compared with most of Norway's trading partners. In the spring of 1977 the Scandinavian currencies were devalued within the snake. A further Swedish devaluation of 10 per cent in the autumn of 1977 made the Norwegian position more vulnerable, and in February 1978 the Norwegian currency was devalued by 8 per cent.

The high currency rate that prevailed between 1973 and 1977 made Norway's imports less expensive, but it also contributed to the decline in the competitive position of Norwegian industry on major markets.

Furthermore, Norway's exports are predominantly raw materials and semi-finished goods, which makes the Norwegian economy highly sensitive to changes in industrial production and investment abroad. The decline in industrial production in the OECD area in 1974 and 1975 weakened demand for Norwegian goods. The economic upsurge in 1976, when the OECD economies grew at slightly more than 5 per cent, affected Norwegian exports positively. Correspondingly, the lower OECD growth rates in 1977 and 1978 have clearly been detrimental. If economic growth rates in the OECD area remain relatively low for a long period of time, the lower demand for traditional Norwegian exports will pose a serious structural problem for the Norwegian economy.

The anticyclical policy to a large extent sheltered Norwegian industry from a deteriorating international situation. But in this way it also distorted the normal market incentives that guard the behaviour of industrial firms, allowing them to avoid adjusting to new market realities. It also created a distorted labour market. Some firms were being paid to hoard labour, while others were crying out for it.[22] In the long run this policy has hurt productivity, which has been stagnant since 1974. In fact Norwegian industry has lost market shares at home and abroad, with imports soaring and traditional exports suffering. In the midst of this, oil production has been delayed considerably.

The delay in oil production is partly due to the unexpected technical and safety problems discussed previously. It is also due to a licensing policy that has been more restrictive than was anticipated. Before 1973, Norwegian intentions were to license rapidly a relatively large part of the area south of 62° N. In June 1972 the Ministry of Industry declared that 212 new blocks would be licensed in the coming years. This was taken as a signal by Norwegian industry that the level of activity on the continental shelf would be high, which stimulated interest in drilling rigs, supply ships and the like.[23] More than six years later, in the autumn of 1978, only 29 blocks had been licensed. The licensing policy had been significantly more restrictive than anticipated because of the oil price rise in 1973, which changed the macro-economic impact of the proposed increase in production. In the oil plan published in the spring of 1974 there were clear signals of a lower level of activity, and a more restrictive licensing policy. It was estimated that only 7–9 drilling rigs and 40–50 supply ships would be needed on the Norwegian continental shelf by 1980.[24] The discovery of Statfjord in 1974 increased reserve estimates and contributed to a stop in licensing for several years. This restrictive policy had a negative effect on Norwegian producers of rigs and supply ships.

The delay in oil revenues and their reduction because of the escalation of production costs reduced Norway's freedom of action in macro-economic policy and made its foreign debt more alarming. In 1970 Norway had no net foreign debts outside shipping. From 1970 to 1974 foreign debts increased because of investment in the petroleum sector. The increase in debts in 1975 and 1976 was essentially due to oil investment and the shipping rescue operation. But in 1977 a third of the foreign borrowing was for purposes of domestic consumption, and in 1978 it represented the bulk of foreign borrowing. Norway's foreign debt passed the $20,000 million mark in the summer of 1978, which corresponded to half of GNP. This is the highest debt ratio ever attained by an OECD country. A large share of the debts are short-term loans that must be renegotiated or replaced after 1980, and in the 1980s the servicing of these foreign debts could be a heavy strain on Norway's economy.

Because of Norway's position as an oil exporter, and because about a third of the foreign debts are linked to oil investment. Norway's indebtedness is less alarming than it would have been in the case of practically any other OECD country. However, the indebtedness can be seen as mortgaging the country's oil reserves and this may limit Norway's freedom in oil policy. The combined debt as of mid-1978 corresponded to the commercial value of about 200 million tonnes of oil, or three years of oil and gas output in the early 1980s. Thus, for financial reasons, the Norwegian government was again to become interested in a higher level of production.

In many ways the Norwegian government appears to have used its oil revenues in a way that is contrary to the 1974 plan recommendations. Instead of a cautious domestic use of oil revenues, there has been liberal use of anticipated oil revenues. But the context was also the opposite from what was expected, and the purpose of spending the oil revenues was to maintain patterns of production and employment instead of changing them. Thus in many ways it was the same approach that in 1973–4 dictated a cautious use of oil revenues. It should also be mentioned that during the 1974–7 period there was significant social progress in Norway. The rise in real wages has already been commented upon, and in addition working hours were reduced, working conditions were substantially improved, and the environment received much better attention. The weak point in this picture is industrial performance, and this might eventually compromise both economic progress and social and political stability.

The delay in oil revenues may be seen as a positive event because it has forced the Norwegian government to revaluate its economic policy. Signals of a new economic policy were evident in the Long Term Programme presented in the spring of 1977.[25] But after the elections in the spring of 1978 a complete reversal was announced in the Supplement to the Long Term Programme.[26] The government stated that there was hardly any room for an increase in private consumption and that strict measures were necessary to restore industrial performance, increase exports and reduce imports. It also announced a ceiling on public expenditure and public investment, but it did feel that private investment should be stimulated. The government declared its intention to take part in wage settlements in order to keep the growth of real wages at or below the prevailing level in other countries. It also announced a more effective price control. The budget of 1979 included new measures of austerity, making 1979 in Norway the equivalent to 1977 in the UK, when real wages were reduced for the benefit of private capital and foreign trade.

The success of the new economic policy after the prices and wages freeze will depend on wages being kept under control, and effective control of prices will be helpful in this respect. But it is questionable to what extent a policy of stagnant real wages is feasible in the Norwegian context of extremely well organised trade unions. Consequently, the new economic policy may require considerably higher unemployment or a more authoritarian incomes policy in which the government determines the wage settlements and curbs the rights of the trade unions at the factory level. Either option would be politically difficult and could easily increase the level of conflict in Norwegian society. In the long run, the success of the policy will hinge on the government's ability to formulate and implement an industrial policy that creates competitive units outside the petroleum sector. Unless this is achieved, the Norwegian economy will tend to become rather one-sided.

Having oil makes the UK and Norwegian economies rather different from those of most other OECD countries as long as energy prices rise. Both countries ought to be in an enviable position. To be West European, industrialised and to have oil should be the best of all economic worlds. In reality, the situation is more complex, with the long-term benefits of petroleum being more doubtful.

In the 1980s the UK has a unique historical opportunity to restore its relative position among the industrial countries. This depends on modernising industry, both plant and structures. There is obviously a danger that oil revenues combined with a tight monetary policy will instead lead to a general inflationary pressure, an increasing exchange rate, declining productive investment and a deterioration of the social climate. In this sense, oil could accelerate the relative industrial decline of the UK.

Norway in the 1980s runs the unique risk of deindustrialising through sudden and easy wealth. Problems with incomes policy and industrial policy present a persistent temptation to solve economic and political problems by increasing the use of oil revenues, and in the longer run relying upon an increasing output of oil and gas. With declining competitiveness in traditional industries, this temptation is likely to become more acute. Indeed, Norway faces a unique dilemma in relation to its oil revenues. If used domestically, they lead to a general cost pressure, compromising industrial competitiveness. If not used domestically, they lead to financial surpluses likely to drive up the exchange rate, also compromising industrial competitiveness. Thus, to some extent there is a conflict between oil revenues and industrial income.

Oil revenues are qualitatively different from other forms of income; they represent a rent. For governments, oil revenues represent easy money. Thus, they can use oil revenues to create a comfortable position for themselves. The problem is, however, that within a complex industrial economy, the ability to absorb a sudden influx of easy money is limited, so that oil revenues tend to become a substitute for other income rather than a supplement. Consequently, the net short-term gain may be less than the large oil revenues indicate in a dynamic perspective; the short-term use of rentier income may compromise the long-term generation of other forms of income. Referring to Gresham's Law: 'Bad money drives out good', it is tempting to try to formulate another one: 'Easy money drives out difficult money; rentier income drives out productive income; petroleum revenues drive out industrial income'.

Oil and Energy Policy

Even prior to finding oil, the UK and Norway were fortunate in being among the Western European countries that were least dependent on energy impacts. UK coal and Norwegian hydro-electricity have provided fairly plentiful domestic energy sources, and have provided the basis for industrial development that is based on fairly abundant energy. The discoveries of oil and gas are part of this historical continuity.

Table 7.3: Structure of Energy Consumption, UK and Norway (mtoe)

	UK		Norway	
	1973	1975	1973	1975
Coal	86 (36%)	70 (34%)	1 (5%)	1 (4%)
Hydro	1	1	10 (52%)	12 (59%)
Nuclear	5 (2%)	8 (4%)	–	–
Gas	25 (11%)	32 (16%)	–	–
Oil	113 (50%)	92 (45%)	8 (42%)	8 (39%)
Total	230	203	19	20

Sources: OECD, Energy Statistics 1973–5 and OECD, Energy Balances 1974–6.

The past abundance of energy has also fostered relatively wasteful patterns of energy consumption. In 1973 total energy consumption *per capita* was 4.1 toe in the UK and 4.9 toe in Norway. In the case of the UK, the high energy consumption is partly explained by coal, which usually gives high total figures for energy consumption because of the large heat losses it creates. But outdated industrial equipment, insufficient building insulation and low gas prices also contribute to the high figures for the UK.

In Norway, the high energy consumption can in part be explained by an important energy-intensive industry based on cheap hydro-electricity. There is also high electricity consumption in the residential and commercial sectors because of the low price of electricity. As a result, Norway has the world's highest electricity consumption *per capita.* However, in both countries the consumption of oil is compara-tively low. In 1973 UK oil consumption *per capita* was 2.0 tonnes and in Norway it was 2.1 tonnes *per capita.* This was less than the French or West German consumption of oil *per capita.* France and Germany are more energy efficient but also much more dependent on oil.

Both Norway and the UK are facing difficult choices in co-ordinating oil and energy policy. There are obvious trade-offs between energy conservation, the expansion of alternative energy production and increasing the use of domestic oil and gas. In the UK, there is a clear potential for energy conservation, but this may be difficult to combine with gradual industrial expansion. The alternatives are either to increase the domestic output of coal and nuclear energy, or to use more oil and gas. With the first alternative there are the obvious social and environmental problems related to the production and consump-tion of coal, and the safety problems and increasing capital costs and fuel costs of nuclear plants. The petroleum alternative implies accelerated depletion of the finite reserves as well as losses in foreign exchange earnings. Because of the economic slump, UK energy consumption fell by 9.5 per cent from 1973 to 1975, and the 1973 level was only reached in 1979. For the future, growth rates of energy consumption of 2.0–2.5 per cent a year may be a reasonable estimate, depending upon the success of conservation, and assuming an economic growth rate of 3.0 per cent a year. On these assumptions the UK may again be a substantial net importer of oil by the turn of the century.

Table 7.4: Alternative UK Energy Futures 1975–2000 (mtoe)

		1975	1980	1985	1990	1995	2000
Total energy	(a)	206	227	251	277	306	338
demand	(b)	206	233	264	299	337	381
Coal output		75	78	81	86	91	96
Nuclear Output		8	13	15	18	38	60
Hydro, etc.		1	1	1	1	1	1
Gas output		32	35	40	43	40	35
Oil output	(c)	0	100	125	115	100	75
	(d)	0	100	100	100	100	100
Oil and gas	(a) (c)	92	0	−11	14	36	71
imports	(a) (d)	92	0	14	29	36	46
	(b) (c)	92	6	2	36	67	114
	(b) (d)	92	6	27	44	67	89

Notes: a. 2% yearly growth in total energy demand.
     b. 2.5% yearly growth in total energy demand.
     c. Accelerated oil profile, production peaking 1985.
     d. Extended oil profile, production kept constant 1980–2000.

Sources: *Energy Policies and Programmes of IEA Countries, 1977 Review* (OECD, Paris, 1978), p. 166, and author's own calculations.

The output of coal has declined in recent years, but can be expected to grow again because of new discoveries. Nuclear power presently accounts for less than 5 per cent of the energy balance. It is expected to yield 15 mtoe by 1985. The maximum likely by the year 2000 is in the order of 60 mtoe.[27] However, such a figure will imply a substantial effort of nuclear development in the 1980s and 1990s. The output of natural gas is building up in the 1970s, reaching 50–55 mtoe by 1980, and it could level off by 1985–90 unless large new fields are found and put into production. The oil production from proven and probable reserves will make the UK self-sufficient around 1980, but production is likely to peak by 1985. Production might be increased, or extended, through future discoveries, but the potential here may be limited. In any case, a peak seems likely by the early or mid-1990s. The reversal to the position of a net importer of oil will then depend upon the success of energy conservation, on economic growth and on the profile of oil production.

In the most favourable of hypotheses, with low growth in energy demand and an extended oil production profile, the UK will remain a modest net importer of oil throughout the 1980s and 1990s. In the least favourable of hypotheses, with a high growth in energy demand and an accelerated oil production profile, UK oil imports will be higher by the year 2000 than they were in 1975. Perhaps the most likely outcome is that of a low growth of energy demand combined with an extended oil production profile, which will make the UK a small net importer of oil by 1990, and then with increasing oil imports through the rest of the century. Given the prospects for a new price rise by the 1980s or 1990s, there should be good arguments for opting for an extended profile of British oil production.

In Norway, price rises and conservation measures contributed to a small increase in energy consumption from 1973 to 1975. The total Norwegian potential for firm hydro-electric power (usable in at least 27 years out of 30) is estimated at 130 TWh, corresponding to 22.3 mtoe.[28] In addition there is an estimated potential of 20 TWh of occasional power. By 1978, 73 TWh of firm power and 11 TWh of occasional power had been developed. To overcome the unreliability of occasional power, imports of 4 TWh were assured through contracts, giving a total assured supply of 77 TWh. The development of another 17 TWh by 1985 has been licensed, which will bring the total assured supply to 94 TWh. Consumption in 1975 was estimated at 77 TWh, but this figure may be too high, as more recent estimates are 73.5 TWh.[29] Growth rates for total energy consumption were previously estimated to be 3.3–3.4 per cent a year.[30] In the light of recent experience, a more realistic estimate is about 2.7–3.2 per cent a year, depending on energy conservation and economic growth rates. This confronts Norway with the choice of using oil to generate electricity or developing nuclear power. The projected expansion of electricity consumption from 73.5 TWh in 1975 to 94 TWh in 1985 implies an average yearly growth of 2.5 per cent. At this rate of increase the available potential of 130 TWh would be developed by 1998. But this development will be politically controversial because of the threat it poses to the environment. In

Table 7.5: Alternative Norwegian Energy Futures (mtoe)

		1975	1980	1985	1990	1995	2000
Energy	(a)	19	22	25	28	32	37
demand	(b)	19	22	26	30	36	42
Electricity demand		11	12	14	16	18	20
Hydro supply		11	12	14	16	16	16
Coal		1	1	1	1	1	1
Excess electricity demand		—	—	—	—	2	4
Therm. plant oil demand		—	—	—	—	3	6
Oil demand	(a)	8	9	10	11	18	26
	(b)	8	9	11	13	22	31
Oil and gas	(c)	9	65	70	55	25	10
output	(d)	9	65	70	55	50	50
	(e)	9	65	70	100	100	100
	(f)	9	65	70	150	150	150
Oil and gas	(a) (c)	1	56	60	44	7	−16
exports	(a) (d)	1	56	60	44	32	24
	(a) (e)	1	56	60	89	82	74
	(a) (f)	1	56	60	139	132	124
	(b) (c)	1	56	59	42	3	−21
	(b) (d)	1	56	59	42	28	19
	(b) (e)	1	56	59	87	78	69
	(b) (f)	1	56	59	137	128	119

Notes: a. 2.7 per cent demand growth per year.
b. 3.2 per cent demand growth per year.
c. No new oilfields put into production.
d. Oil and gas production at 50 mtoe in the 1990s.
e. Oil and gas production at 100 mtoe in the 1990s.
f. Oil and gas production at 150 mtoe in the 1990s.

Sources: For the base year 1975 OECD Energy Balances 1974–6 are used, adjusted for hydro. Assumptions are that coal consumption will remain at 1 mtoe a year, and that hydro power development will stop in 1990.

practice, the ceiling for hydro-electric development is likely to be reached earlier, perhaps by 1990, at a level of 106 TWh. To generate the equivalent of 1 TWh in an oil-powered thermal plant requires from 200,000 to 250,000 tonnes of oil. The result is a quickly increasing domestic demand for oil in Norway in the 1990s if nuclear power is not developed.

In the worst case, if no new oil and gas fields are developed and no nuclear power is developed, Norway's position as an exporter of

petroleum will be confined primarily to the 1980s. Considerable
imports of petroleum are likely by the year 2000 even if conservation is
successful and economic growth rates are low. Alternatively, if new
fields of oil and gas are developed and the ceiling of 100 million tonnes
of oil equivalents is reached or production reaches an even higher level,
then Norway's freedom of action in energy policy will be considerable.
A more vigorous nuclear alternative has not been considered because
no nuclear power plants in Norway could be operational before 1990.
If a decision to develop nuclear power is made around 1980, there
might be a considerable nuclear power supply by the early 1990s which
would reduce domestic demand for oil, allowing more petroleum exports
or giving Norway more freedom in its depletion policy for oil and gas. In
any case, Norway faces a delicate political choice, and there is a risk that
political and administrative factors may compromise the actual exploitation of Norway's naturally competitive advantage from abundant
energy.

Oil and Industrial Policy

The UK and Norway both have serious industrial problems which are
in part structural, in part related to management methods, and in part
related to labour. These problems mean that the potential industrial
benefit from oil is not fully exploited. But oil and oil revenues also
allow governments an improved basis for industrial policies that can
correct these problems.

The structural problem in UK industry largely involves outdated
capital equipment which undermines productivity and industrial
innovation. In Norwegian industry the structural problem is the
decentralisation of capital. There are a large number of small enterprises, and even though their capital equipment is generally modern,
the units are often too small to be competitive and make full use of
the capital. The management problem, in both the UK and Norway, is
primarily due to insufficient education in business administration. In
the case of the UK there is also the problem of antiquated social
attitudes, whereas in Norway there is the additional problem that
management is more often linked to the productive function than to
marketing.

The labour problem in the context of the UK is that the trade union
movement is generally organised along craft lines, with many different
unions competing within one firm. In Norway the problem is a

generally low mobility of labour and a resistance to change that is often shared by management.

In the face of these problems, the discovery of oil provides a basis for industrial improvement at three levels. It provides the domestic raw materials for the oil refining and petrochemical industries; it creates a domestic market for new products and services; and it provides capital that can be used for industrial investment, reorganisation and restructuring.

Historically, the UK had a large refining and petrochemical industry, while Norway did not, and was traditionally a net exporter of refined oil and petrochemical products. Norway's first refineries were only built in the early 1960s. Prior to the oil crisis, both countries had plans for substantially increasing their oil refining and petrochemical industries. Since 1974 there has been a large excess capacity in Western European refining. Some of the old refineries in the UK have been scrapped, and in both countries plans for new construction have been scaled down or shelved. Domestic oil does, however, give the UK and Norway a competitive advantage in petrochemicals, and as a result both are embarking on new projects.

The need for equipment to exploit the oil of the continental shelf is an obvious incentive to the expansion of domestic industries into these areas. As more and more world oil production takes place on continental shelves, the emerging domestic industry will have a basis for specialisation and an international market. The new market for offshore drilling equipment can be divided into two segments, the traditional products of heavy engineering and new technology. Oil operations offshore require both traditional products such as steel structures, cranes, valves, pumps and generators, and new products such as undersea pipelines, drilling rigs, production platforms, supply ships and eventually new kinds of offshore production systems.

UK industry has been fairly good at moving into the traditional products market, but it has largely missed the market for new products.[31] Norwegian industry, by contrast, has been successful in the new products market,[32] but did to a certain extent miss the traditional products market. Though UK industry was able to cope with new demand for traditional products which required little modification of production patterns,[33] its failure to move into the market for new products came from a reluctance to take risks with new technology management. In addition, UK shipyards were suffering from particularly bad industrial relations until the mid-1970s. Norwegian industry was able to adapt to the new technologies and cope with the demand for new products by using its technological experience with shipbuilding. Within a short time Norwegian shipyards scored successes in mobile drilling rigs and supply ships.[34] A rig of Norwegian design, the H-3, became the world's best-selling rig, and by 1975 its company was the world's second-biggest rig builder, controlling about 20 per cent of the offshore rig market.[35] This flexibility was caused by a willingness to take risks that was due to a record of high investment and the gambling attitude of Norwegian shipowners.

The above description is rather an over-simplification. In January 1973 the UK government created the Offshore Supplies Office (OSO) in order to co-ordinate and stimulate deliveries of UK-produced goods and services. According to a government decision of 1975, the OSO can inspect offshore operators and ensure that they use UK goods and services that are competitive. As a result, the share of UK goods and services in North Sea oil production has increased considerably. The OSO can demand that the deadline for tenders be delayed, and can then take the initiative to organise an acceptable UK offer. The OSO also plays a more active part by identifying areas in which UK industry is uncompetitive, then encouraging local firms to develop the competence required. The OSO also has an important planning and development function. It is actively engaged in offshore industrial research and development and bids on offshore projects abroad. It also tries to offset through exports the fluctuations in demand from the UK continental shelf alone. Finally, the OSO can recommend public financial support to firms engaged in deliveries for offshore oil production. Because of subsidies UK rigs and platforms are highly competitive on the UK continental shelf. From 1974 to 1977 UK deliveries to their part of the North Sea rose from 40 to 62 per cent.[36]

In Norway, the role of domestically produced goods and services in offshore oil exploitation is increasing, mainly because of Statoil's participation at Statfjord. There have been some remarkable initial successes, such as the H-3 drilling rig and the concrete production platform Condeep.[37] Since 1969 there has been a clause in the production licences that has required companies to use Norwegian goods and services when they are competitive.[38] Because of the small corporate units in Norwegian industry and their limited ability to co-operate, Norwegian industry has been unable to win a larger share of the market. Norwegian firms tend to compete in the same limited market, confining their efforts to areas in which other firms have already scored successes. In recent years, Norwegian firms have increasingly co-operated with foreign partners in order to gain expertise and technology,[39] but this has not been a complete success. Perhaps Norwegian firms should concentrate on creating larger productive units. For example, the success of the concrete platform is in part due to co-operation among all the relevant contractors.

The government's desire to stimulate the use of Norwegian goods and services in oil production has not been followed up administratively. There is as yet no separate supplies board for the offshore oil industry. Instead, Statoil is supposed to defend Norwegian industrial interests and give preference to Norwegian goods and services. This policy is bound to be ineffective because it delegates an essentially administrative task to a commercial enterprise. The basic problem is structural, and it is doubtful that Norwegian industry will be able to restructure itself into larger, more competitive units without active state intervention.

As mentioned above, the UK Labour government gave high priority to investment in the industrial sector as a whole in its use of North Sea oil revenue.[40] The purpose of increasing the level of industrial investment was to raise output and productivity. The attainment of this goal also required a qualititative improvement in management that is combined with harmonious labour relations. The Labour government clearly intended to intervene more than in the past by supplementing the private capital markets with public funds, particularly in cases of high-risk or long-term projects. It also intended to use the National Enterprise Board and the regional development agencies to promote investment in government-sponsored projects and in projects sponsored jointly with private industry. Planning and development agreements were thought to be an important tool in developing new products and technologies. In addition, the government's industrial strategy envisaged more industrial democracy in both the public and the private sector through the institutionalisation of worker co-operation and participation in management. The aim here was to improve labour relations and create an industrial climate that can stimulate technological innovation and productivity gains.

In Norway, with its fears of an overheated economy, the government aimed at reducing employment in manufacturing. The transfer of labour from industries that are exposed to foreign competition to industries that are sheltered from foreign competition is essentially a transfer from manufacturing to the service sector. In the spring of 1975 the Norwegian government presented a plan for the gradual restructuring of industry.[41] The goal was both to create larger units and to keep a network of smaller enterprises. This plan also recognised

the need for industrial employment to decline because of the
government's anticyclical policy. The major objective of the anticyclical
policy was to maintain employment levels through a given period, but
this in many ways made the situation worse by intensifying the wage
cost pressures by reducing the mobility of labour and capital, and
generally distorting the signals from the market to industry.

By 1978 it was clear that the anticyclical policy could no longer be
maintained, and the Norwegian government was in acute need of an
industrial policy. At the same time, Sweden's largest industrial firm —
the Volvo car manufacturer — needed capital desperately; in the spring
of 1978 an agreement between Volvo and Norway was announced. A
new holding company was to be created, with 40 per cent Norwegian
capital. Volvo was to create several thousand jobs in manufacturing in
Norway, and in return a new subsidiary, Volvo Petroleum, was to get
access to Norway's continental shelf. At this point the Norwegian
government also declared that the ability of a company to help in
Norway's industrial development and create jobs in Norway would be
one criterion for granting licences on the continental shelf.

The Volvo deal signified a complete turn-around both in industrial
policy and in oil policy. The creation of manufacturing jobs is now
explicitly seen as positive, while it was seen as negative only a few
years earlier. To link two different policies in one large agreement was
also a new practice for Norway. The Volvo deal was clearly precipitated
by Volvo's need for capital and Norway's need for jobs, and it does not
seem to have been the result of a long-term strategy by either party.
For Norway, the link between oil and industrial policy was a sign of an
inadequate industrial policy. With Norway's previous economic
position it should have been unnecessary to use oil directly as a means
of securing industrial employment. Linking industrial employment to
access to oil compromises oil policy by pushing up the level of oil
production and creating more foreign participation. A policy for the
planned restructuring of Norwegian firms, aiming at larger units
that are internationally competitive, might be more useful in creating
stable employment and less harmful to Norway's freedom of action in
oil policy.

Social and Regional Aspects

When a new economic activity that is highly capital-intensive is rapidly
introduced into a society, the social impact can be fairly dramatic. In

the case of the UK, the size of the new activity relative to the existing economy is such that the social change is essentially a regional problem. In Norway the relative size of the new activity is such that the whole nation is affected, and significant changes in the pattern of production, employment and residence are almost inevitable. These changes can in turn be very disadvantageous to individual activities.

The gas activities off south-east England have had little social or regional impact. But the oil and gas activities further north, off northern England and Scotland, have had a stronger impact, particularly in Scotland. There the new oil industry has created local inflation through rising property prices, pressure on local communities and demand pressure on the local industrial work-force.[42] Given that Scotland's productive factors were far from fully utilised before the advent of the oil industry, this local demand pressure surely has advantages.[43] In fact, eastern Scotland has changed from one of the economically most depressed areas of the UK into one of the more prosperous ones, and this has had a positive effect on the whole Scottish economy.

The fact that UK oil revenues are used all over the UK reduces the risk of local economic overheating. The problem is that the impact of oil may not last long, because depletion policy is made according to UK needs rather than to local Scottish needs. Thus the demand pressure created by oil does not promote a lasting structural change in the Scottish economy.[44]

A depletion policy set according to Scottish needs would have a more moderate rate of development, and a much longer profile of production. A Scottish depletion policy would make the initial problems of adaptation smaller and the positive effects more long lasting as well as enhancing the possibility of carrying out a structural change in the Scottish economy. This consideration is an important motivation behind Scottish nationalism.[45] It is believed that with Scottish control of depletion policy and oil revenues, Scotland would have a strong economy and good incentives for expansion, thus breaking with the depressed economic situation of the past.[46]

The response of the UK government has been to move the head-quarters of BNOC and the Offshore Supply Office to Scotland, and to give certain preferences to Scottish goods and services. Even if Scotland remains part of the United Kingdom, and does not get control of the depletion policy and the oil revenues, the situation is not altogether bleak. As stated above, the UK economy might well move to a lower rate of depletion, and there is the possibility of new discoveries off western Scotland that might extend the time horizon for production considerably.

In Norway there are certain parallels to the Scottish situation, with the difference being that at the advent of oil production the Norwegian economy was not depressed. The development of the oil industry in Norway necessarily creates changes in the patterns of production, employment and settlement, and thus produces some disadvantages at the level of the individual. The oil industry also leads to deteriorating conditions for certain businesses and sectors of the economy. On the whole, the oil industry in Norway has met with considerable scepticism and opposition, not only from the radical or ecological left but also from established businesses and economic sectors. This explains the cautious attitude taken by the Norwegian government from the outset.

The direct effects of the Norwegian oil industry have so far been concentrated in the Stavanger area in south-west Norway and a few other places where rigs and platforms have been built. The consequences for the labour market and the housing situation are particularly visible. In affected areas there is economic overheating and wages are considerably higher than the national average. Marginal groups of workers are employed more easily, but the cost pressure creates serious problems for local industries not linked to oil. The acute housing shortage is a serious social problem.[47] There are other social problems caused by the rapid local economic upsurge that include negative effects on family relations and youth socialisation. Some of these problems appear modest by international comparisons, but in the traditionally stable Norwegian or Scottish social context they are seen as serious and highly undesirable.

Working conditions offshore merit special mention. The operating companies in the UK and Norwegian sectors of the North Sea have traditionally been foreign, mostly American, and have usually employed a high proportion of foreigners. Working conditions, social and health conditions and even safety standards in the international oil industry are usually lower than in the UK or Norway. The high priority given to development and production in the North Sea has naturally led to a low priority being given to the safety working conditions and the social and health care of employees.[48] In addition, there have been numerous obstacles to unionisation among them because of the high number of foreigners. Consequently, working and social conditions in the North Sea still leave much to be desired from a UK or Norwegian point of view, and this is a permanent source of friction between UK or Norwegian labour and the foreign oil companies. Operating companies based in the two countries are probably more inclined to conform to established local practices, and they are also more exposed to public pressure.

Protection of the Environment

The offshore oil industry is a permanent threat to the environment because of the potential for spills and blow-outs. This is also a problem of economic policy because there is a trade-off between economic interests related to oil and economic interests related to the environment, mainly fishing. In the North Sea this issue is more serious than, for example, in the Gulf of Mexico, because the lower temperature of the water means that oil particles are broken down more slowly, and because of the fishing industry, which makes the North Sea very important to Europe's food supply. In the Norwegian Sea, off northern Norway, this issue is even more important, because of lower water temperatures and higher catches of fish.

So far little is known about the exact ecological effects of oil pollution,[49] but experiments indicate that oil has a fairly lethal effect upon plankton, fish eggs, larvae and shellfish. Because oil is lighter than water, it spreads over a large area quite quickly and can be taken by currents towards the large fishing areas off northern Norway. A very large oil spill might reduce the fish stock for many years, and it could induce the fish to seek new breeding grounds elsewhere.

Fishing is very important to the Norwegian economy, particularly for exports and to the towns along the Norwegian coast. The political weight of the fisherman and the coastal population explains why Norway has given a high priority to the protection of the environment ever since oil production started. Norway was the first North Sea country to introduce requirements for oil companies to avoid pollution and damage to marine life.[50] It also established an Oil Defence Council with fairly wide powers. The UK later, partly under the auspices of European Community rules, introduced stricter legislation in this field.

Given the increasing level of oil activity in the North Sea, there is an increasing risk of pollution, especially through blow-outs. The only serious blow-out in the North Sea took place at Ekofisk in Norwegian waters in April 1977. The blow-out lasted for eight days, letting about 22,500 tonnes of oil out into the sea. A Royal Commission has established that the accident was essentially due to human error rather than technical faults, which were of little significance. It deemed that the responsibility for the accident belongs to the administrative and organisational systems that were not prepared to correct errors and react with counter-measures.[51] The Norwegian Petroleum Directorate was explicitly criticised for neglecting its supervisory duties. Furthermore, the Ministry of Environment and the State Pollution Control Agency were not prepared for the accident, which was eventually stopped through the assistance of a Texan expert.

The Ekofisk accident was relatively small. The platform did not catch fire and the weather conditions were favourable. If the accident had taken place in winter, or if the platform had caught fire, the blow-out could have reached serious proportions. Greater quantities of oil would have been discharged into the sea, and stopping the blow-out might have required drilling an additional well to relieve the pressure. This would certainly have taken months to accomplish. In the aftermath of the blow-out, the relevant Norwegian administration systems are supposed to be reinforced, but only a new accident can test whether or not they have been improved.

The political effect of the blow-out was to postpone licensing in the southern waters and delay drilling in the northern waters for several years. The risk of blow-outs and pollution is an excellent argument in favour of a moderate rate of development in the oil industry. It also seems that oil companies based in the countries involved would be more sensitive to public concern for the environment. In the Ekofisk case, the foreign operator had not adequately prepared the personnel, and both the qualifications of the personnel involved and the company's organisation were inadequate.[52] Similarly, at the Alexander Kielland accident in March 1980, a large part of the personnel had not been trained in safety precautions. This probably was a direct cause for the high casualty figure.[53] In general, foreign operators appear to give a fairly low priority to safety.[54]

Notes

1. Edward F. Denison, 'Economic Growth' in Richard E. Caves (ed.), *Britain's Economic Prospects* (George Allen and Unwin, London, 1968), pp. 231–78.

2. *The Challenge of North Sea Oil* (Her Majesty's Stationery Office, London, 1977), p. 5.

3. Introduction to Caves, *Britain's Economic Prospects*, p. 5.

4. Stephen Blank, 'Britain's Economic Problems: Lies and Damn Lies', in Isaac Kramnick (ed.), *Is Britain Dying? Perspectives on the Current Crisis* (Cornell University Press, London, 1979), pp. 66–88.

5. Ibid., p. 68.

6. Sheila Page, 'The Value and Distribution of Benefits of North Sea Oil and Gas, 1970–1985', *National Institute Economic Review* (1978), pp. 41–58.

7. *The Economist*, 9 September 1978, p. 15.

8. *National Income and Expenditure 1966–1976* (Central Statistical Office, London, 1977), p. 17.

9. *The Economist*, 9 September 1978, p. 15.

10. Ibid., p. 47.

11. Ibid., p. 48.

12. Ibid., p. 57.

13. *The Challenge of North Sea Oil* (Her Majesty's Stationery Office, London, 1977), pp. 9 ff.

14. Ibid., p. 11.

15. *Langtidsprogrammet 1973–1977*, St.meld.nr. 71 (1972–3), (Ministry of Finance, Oslo, 1973), p. 327.

16. Ibid., p. 88.

17. *Petroleumsvirksomhetens plass i det norske samfunn*, St.meld.nr. 25 (1973–4) (Ministry of Finance, Oslo, 1974).

18. *Naturressurser og økonomisk utvikling*, St.meld.nr. 50 (1974–5), (Ministry of Finance, Oslo, 1975).

19. *The Economist*, 26 July 1975, 'North Sea Oil Survey', p. 6.

20. Ibid., p. 26.

21. *The Economist*, 15 November 1975, 'The Next Richest Nation', p. 7.

22. *The Economist*, 1 April 1978, p. 82.

23. Sigmund Gjesdahl, 'Er vi "blå-øyde arabere" ', *Dagbladet*, 18 September 1978, p. 3.

24. *Petroleumsvirksomhetens plass i det norske samfunn*, Supplement, pp. 18 ff.

25. *Langtidsprogrammet 1978–1981*, St.meld.nr. 75 (1976–7), (Ministry of Finance, Oslo, 1977).

26. *Tillegg til Langtidsprogrammet 1978–81*, St.meld.nr. 76 (1977–8), (Ministry of Finance, Oslo, 1978).

27. Carroll L. Wilson (ed.), *Energy: Global Prospects 1985–2000* (McGraw-Hill, New York, 1977), p. 203.

28. In Norway a conversion factor of 1 TWh = 0.15 mtoe is normally used.

29. *Langtidsprogrammet 1978–1981*, p. 81.

30. *Energiforsyningen i Norge fremtiden*, St.meld.nr. 100 (1973–4), (Ministry of Industry, Oslo, 1974), p. 40.

31. *The Economist*, 26 July 1975, 'North Sea Oil Survey', p. 21.

32. Ibid., p. 34.

33. Ibid., p. 21.

34. *Operations on the Norwegian Continental Shelf*, Report No. 30 to the Norwegian Storting (1973–4), (Ministry of Industry, Oslo, 1974), p. 54.

35. 'North Sea Oil Survey', p. 34.

36. *Aftenposten*, 18 September 1978, p. 26.

37. *Norsk Oljerevy*, no. 6 (1978), p. 8.

38. *Operations*, p. 54.

39. *Norsk Oljerevy*, no. 6 (1978), p. 8.

40. *The Challenge of North Sea Oil*, pp. 8 ff.

41. *Norsk Industries Utvikling og Fremtid*, St.meld.nr. 67 (1974–5), (Ministry of Industry, Oslo, 1975).

42. David Taylor, 'The Social Impact of Oil' in Gordon Brown (ed.), *The Red Book on Scotland* (EUSPB, Nottingham, 1975) pp. 270–81.

43. D. I. Mackay and G. A. Mackay, *The Political Economy of North Sea Oil* (Martin Robertson, London, 1975), pp. 111 ff.

44. Ibid., p. 136.
45. Ibid., p. 165.
46. Ibid., p. 185.
47. *Sosiale og helsemessige konsekvenser av petroleumsvirksomheten* (NOU, 1975: 38; Ministry of Social Affairs, Oslo, 1975), pp. 10 f.
48. Ibid., p. 18.
49. Knut Bryn (ed.), *Oljen og det norske samfunnet* (SNM, Trondheim, 1974), p. 110.
50. Patricia W. Birnie, 'The Legal Background to North Sea Oil and Gas Development' in Martin Saeter and Ian Smart (eds.), *The Political Implications of North Sea Oil and Gas* (Universitetsforlaget, Oslo, 1975), pp. 19–50.
51. *Ukontrollert utblåsing på Bravo 22. april 1977* (NOU, 1977: 47; Ministry of Justice, Oslo, 1977), p. 60.
52. Ibid., p. 59.
53. Bernt Eggen and Håkon Gundersen, 'Oljepolitikk på Dispensasjon in *Nordsjotragedien*, pp. 156–71.
54. Bjorn Nilsen and Bernt Eggen, *Det Brutale Oljeeventyret* (Oslo, October 1979), pp. 14 ff.

8 OIL AND POLITICAL ECONOMY

The Relevance of Oil

The question of how oil affects the relationship between social classes in UK and Norwegian societies has so far not been debated much, but it is nevertheless of great importance. When a new capital-intensive economic activity such as the oil industry is rapidly introduced into a social context, it affects the already existing balances and relationships. In this respect, the pattern of organisation and the control of the oil industry means to a large extent control over one of the most dynamic elements in the UK and Norwegian economies over the next years, and it also means control over an important market for industrial goods. In addition, control of the oil industry and the revenues from oil means control over a significant share of the money circulating in the two economies.

The development of the North Sea oil industry meant a historical change for UK and Norwegian capitalism, leading to a rationalisation of structures. Left to private initiative – even without massive foreign participation – the risks involved in petroleum exploration and development would have eliminated less competitive firms and left a select number of efficient and successful firms, which could then use their strong position in the oil industry to expand into other areas of the national economies. The result would have been a restructuring of UK and Norwegian capitalism on the basis of larger and more capital-intensive units. This prospect was evidently contrary to the interests of most of the business community in the two countries. Consequently, since the discovery of oil there has been a split in UK and Norwegian business, with some capitalists hoping to participate directly in the North Sea bonanza and use the profits as a means of further expansion, and others fearing the impact of North Sea oil on their industries.

This has to a large extent been a split between a group composed of the financial establishment and the industries participating in oil development, and a group representing the bulk of manufacturing not related to oil. In Norway the banks and the shipping interests were eager to take part in the oil industry, while the majority of manufacturers were sceptical about it.[1] This divided attitude among capitalists prevented the non-socialist parties in Norway from defending

211

wholeheartedly the virtues of private enterprise and economic liberalism where the continental shelf was concerned. It also made it fairly easy for Labour governments in the two countries to opt for a system of extensive state control, and even state participation and state owner- ship. This policy probably had the implicit support of the majority of capitalists, who saw state control as a defence of their own position. In addition, state control could be seen by fairly large parts of UK and Norwegian industry as a bulwark against excessive foreign dominance, and as a means for attaining greater domestic involvement in producing goods and services for the oil industry.

Against this background, the North Sea model not only appears to be a radical social democratic innovation in resource management, but also seems to be a defensive move by certain private interests against a potential threat from other private interests, both domestic and foreign. From this perspective the North Sea model can be seen as a distortion of UK and Norwegian markets in order to moderate the restructuring of the business community that otherwise would have been extensive. This point of view is particularly relevant in the Norwegian context, but it is also relevant to the UK, given the weak financial position of large parts of UK capitalism. Thus, to a certain extent the North Sea model is a tacit alliance between the weaker elements of business and the Labour movement under the management of state technocrats. But the North Sea model also implies a tacit alliance between elements of the bureaucracy and the parts of the private sector that are involved in the oil industry. The North Sea model changes the political economy of the UK and Norway by giving the state a direct stake in the accumulation of capital, and giving it a more central role in the pattern of economic organisation. This means a fundamental change in the balance between the public and private sectors.

The Background of the Balance between the Public and Private Sectors

Even before the advent of oil, the UK and Norway had fairly large public enterprises. In the UK the major nationalised industries were the utilities, the railways, the coal industry and the steel industry. In Norway, there have traditionally been fewer nationalised enterprises, the most important ones being steel, munitions and arms manufacturing, the railways, iron mining and the bulk of the aluminium industry. Some of the state-owned firms were the result of an active nationalisation policy designed to gain more influence on economic life. In the UK public ownership of the coal and steel industries reflects such a strategy, while in Norway public ownership of the steel and the aluminium industries represents the same trend.

In recent years state ownership has grown in both countries, partly as a result of an active nationalisation policy, but just as often as a result of passive nationalisations in which the state had to rescue bankrupt private firms in order to save employment and technology. Thus, in the UK, public ownership has been actively extended to oil production, while passive nationalisation applied to Rolls-Royce and the shipyards. In Norway, public ownership was actively extended to oil production, to the chemical industry and to the rest of the aluminium industry, while coal production in Spitzbergen and the main parts of the electronics industry were subject to passive nationalisation.

In 1975, before oil production had reached a significant level in either country, firms in public ownership represented slightly more than 10 per cent of GNP in the two countries. The relative importance of gross investment as compared to employment is remarkable in the case of the UK and indicates that nationalised firms are more capital-intensive than private firms. Nationalised firms such as British Steel and Norsk Hydro dominate their domestic markets but compete internationally. In Norway, several of the nationalised firms are primarily export-oriented, such as the aluminium and chemical industries. This explains the large role of nationalised firms in Norwegian exports, 21.5 per cent in 1975. The generally high levels of investment in nationalised firms indicate the active role of technological development. This is the case, for example, with the UK steel and energy industries, and with the Norwegian aluminium, chemical and mechanical industries.

Table 8.1: Importance of Nationalised Firms, UK and Norway, 1975 (per cent)

Turnover	11.0	10.5
Employment	8.0	8.9
Investment	19.0	11.4

Sources: *A Study of UK Nationalised Industries* (National Economic Development Office, London, 1976), p. 13, and *Norwegian Industrial Statistics 1975* (Central Bureau of Statistics, Oslo, 1977), Table 13.

Increased UK and Norwegian public sector involvement in the oil industry has taken two forms:

a direct increase in the public industrial sector, through the active
participation of the state in the oil industry;
an indirect increase in public industrial influence as private firms
have become dependent on orders from the nationalised oil
companies, and as the state has had to rescue or subsidise firms in
the rest of the economy.

Through minimum state participation of 50–51 per cent in new oil
fields in both countries, state participation in the economy increases as
a direct function of the build-up of oil production. In addition, state
participation in the economy increases because of the involvement of
semi-public oil companies such as BP and Norsk Hydro. Thus the public
sector in the economy grows along with the petroleum sector.[2] Within
a short time BNOC and Statoil are likely to become the leading firms
in the UK and Norwegian economies in terms of revenue, profits and
investment, and perhaps exports as well. BNOC and Statoil will also be
major consumers of UK and Norwegian mechanical and engineering
products, and, combined with BP and Norsk Hydro, the public oil
sector could make up the bulk of this market.

   Oil creates particular economic conditions in the UK and Norway.
That accentuates the general problem of adaptation and restructuring.
For example, even if real wages were temporarily reduced, in the UK in
1977 and in Norway in 1979, the combination of oil revenues and
strong trade unions is likely to make real wages increase again. To the
extent that this real wage growth exceeds the level of wage growth in
other countries, the return on investment in most sectors outside oil
will suffer compared with most other countries. This could become a
fairly general and permanent problem for UK and Norwegian industry.
The end result would be a trend towards bankruptcies, mergers and
demands for selective public intervention to preserve employment and
technology. This again creates an increasing need for a clearly defined
public industrial strategy. Fortunately, oil revenues put the state in a
better position to accomplish such a task.

Private and Public Power

In addition to the quantitative growth of the public sector in the UK
and Norwegian economies, there is an important qualitative aspect: the
state becomes the leading business operator and the leading accumulator of capital. Because the oil industry is the most expansive and

propulsive element in the two economies, and because of its high
capital intensity and high level of profits, its ownership and control
have a further political significance. Through control of the oil, the
governments improve their position of command in relation to the
economies and in relation to private capital.

In the immediate post-war period there was an active policy of
nationalisation in several Western European countries, aiming at state
control of the most expansive and propulsive elements in the economy,
to bring certain key profit centres under public control, and to make
the state an important accumulator of capital. Examples outside the
UK and Norway are the Renault factories in France, the IRI in Italy,
the ÖAG in Austria, etc. Later, this policy was eroded, because the new
public firms were in sectors with declining profits such as coal,
because profitable operations were transferred to private enterprise,
and because the public firms had to assume many unprofitable tasks
for social and political reasons. Thus, what was intended to be a
socialisation of profits in many cases turned out to be a socialisation of
losses. This trend was reinforced in the 1970s, when many European
governments had to initiate large-scale rescue operations. Examples are
French and Swedish steel and shipbuilding, etc. In this way the public
industrial sector is extended, but with deficit operations, liabilities
and employment problems. The deficits of public firms imply a
subsidisation of their markets and a transfer of income from the
public to the private sector. This reconfirms the orthodox Marxist
notion of the role of the state in a capitalist economy as securing
private profits and the conditions of reproduction for private capital,
for example by taking over necessary but unprofitable functions.[3]
However, in Western Europe since 1945 public enterprise has not been
synonymous with a socialisation of losses. Examples to the contrary
are the French Renault factories, the French and Italian oil industries,
the Norwegian aluminium industry, the Dutch chemical industry, etc.
The profitability of public firms has mostly been related to the overall
performance of their national economies;[4] the low profitability of UK
public enterprise in the post-war period can to a large extent be
explained by the overall performance of the UK economy. So there is
in Western Europe also a trend towards an industrially active state,
engaging in profitable business operations that accumulate capital. This
is contrary to the orthodox Marxist notion of the role of the state in a
capitalist economy. It should be emphasised that the extension of the
public industrial sector also takes place under non-socialist governments, mostly as a defensive measure, but there are also examples, as in

the case of Norsk Hydro, of the contrary.

In the UK and Norway the creation of large state oil companies with a direct stake in oil production indicates that public profit centres are being built up in the most dynamic sector of the economy, which is aiming at a considerably higher rate of capital accumulation than is common elsewhere in the two economies. To the extent that public industrial policies are successful in creating profitable public firms, there is a likelihood that several of the main profit centres in the economy will belong to the public sector. On the basis of oil alone, at least one part of the public industrial sector will be more profitable than the rest of UK and Norwegian industry. Capital accumulation will take place at a higher rate in the public industrial sector than in private industry. This is something new in the UK and Norwegian economies, and to a large extent in Western Europe.

A more important innovation is linked to the fact that, through the channelling of a large part of the oil revenues to the UK and Norwegian Treasuries, the state at the macro-economic level also becomes an important accumulator of capital. The accumulation of capital by public enterprises, even at a high rate, does not break with the usual practice of capitalist economies to accumulate capital in enterprises at the micro-economic level. With public enterprises, the control of the accumulation process is simply switched from private capitalists to public enterprise managers, who take into account the same basic considerations and aim at profitability.

The accumulation of capital at the macro-economic level by the state introduces a new dimension in the capitalist economy, giving politicians and the bureaucracy control over part of the accumulation process. This clearly improves the macro-economic freedom of the state. It also strengthens the state and the bureaucracy in relation to private or public micro-economic interest, which creates the potential for emphasising macro-economic and social considerations over concerns for micro-economic profitability. This reinforcement of the role of the state is particularly strong in Norway, given the dimension of the oil sector in the national economy. In the UK the same process takes place, but is less significant.

The rise of the public sector contrasts with developments in the private sector. In the mid-1970s, investment income and private profits have in general been negatively affected in both countries. In UK industry, there is a declining trend in profits throughout this century.[5] However, this general trend may obscure important variations over time, and especially between types of industries. For example, there is a possibility that the profit squeeze has hit domestic-oriented industry hardest, while enterprises of an international character have been able to realise higher profits.[6] In any case, there are clear signs of a decline in profits in the late 1960s and mid-1970s, interrupted by increased profits during the reflation of the UK economy under the Conservative government. An important reason for the decline may be the increasing effectiveness of trade unions, which has led to a gradual redistribution of the national income in favour of wage-earners and at the expense of profits and investment income.[7] This explanation contrasts with traditional Marxist theory, which sees capitalist economies creating crises of overproduction and underconsumption.[8] Instead, trade union effectiveness can be seen as creating crises of overconsumption and underproduction, leading to a vicious circle of declining profits and declining investment.[9]

Another reason for the decline in profits may be past macro-economic policy, which gave a higher priority to the UK's role in the world economy than to domestic growth. This led to the stop-go syndrome that impeded investment and gradually led to an outdated capital stock. The performance of other Western countries that followed a macro-economic policy emphasising the positive effects for investment of private profits and real wages suggests that the key variable is the macro-economic policy. Increased trade union militancy can to a large extent be seen as a result of a given macro-economic context. In any case, the largely outdated capital stock meant that the bulk of UK industry was badly prepared to meet the substantial wage increases of 1974 and 1975. In several industries profits declined to a point where massive bankruptcies seemed likely. The loan from the International Monetary Fund in 1976 not only appears as a necessary influx of capital, but also as a useful pretext to keep wages under control, in order to avoid either a further collapse of private profits or more authoritarian measures to restore them. As already mentioned, in 1976 and 1977 the distribution of the national income was altered again, to the benefit of capital at the cost of labour. In 1978 investment seemed to recover quite significantly, but industrial production showed only a minor increase.[10] It is questionable to what extent it will be possible to keep wage increases at a level where private investment and industrial production will be stimulated in a more durable way.

In Norway, profits appear to have been fairly stable in the post-war period, but they were generally increasing in the late 1960s and early 1970s. Norwegian trade unions are comparatively effective, particularly

at the grass roots level in the individual enterprises, accounting for a
substantial wages drift in addition to the general wages settlements.
This has led to a comparatively even distribution of income, but not
eroding private profits to the point where investment decreases. An
important reason for this has been the historical macro-economic
policy, giving a high priority to economic growth and encouraging
investment. This has made Norwegian capitalism able to absorb high
wages without compromising the growth potential, at least until the
mid-1970s. The rapid growth of real wages in the mid-1970s affected
profits in Norway in two different ways. In general profits increased,
in both absolute and relative terms, in industries sheltered from foreign
competition. In industries competing exclusively in Norway there were
significant variations between sectors, but in general profits rose at a
high rate in 1975 and 1976, levelling off in 1977. On the average
profits declined, in absolute and relative terms, in those industries
exposed to foreign competition. In industries competing in the world
market profits have decreased significantly since 1975, and according
to some estimates they were in 1977 only one-third of their 1974
level.[11]

The traditional Norwegian export industries such as shipbuilding
were particularly hard hit by the profits squeeze, and the tremendous
significance of these traditional industries has reduced the average
profitability of Norwegian industry considered as a whole, excluding
oil. According to Norway's Central Bureau of Statistics, the total
profitability of Norwegian industry was 3.3 per cent in 1975, but it has
been claimed that in 1975 total profitability was negative if inflation
is taken into account.[12] Investment declined in 1978 in several
traditional branches of industry, and this trend continued in 1979.

The decline in profits is particularly serious in shipping. Even with
the government-financed rescue operation, part of the Norwegian fleet
had to be sold abroad, and a substantial number of Norwegian shipowners are either bankrupt or close to it. It is doubtful that Norwegian
shipping will ever regain its past position, given increasing international
competition and the modest growth of international trade. This means
that the shipowners, the political hard core of Norwegian capitalism,
have been virtually eliminated. They have moved from being a
principal Norwegian source of foreign exchange with considerable
political influence to being a major burden on the economy and an
object of public policy. The recent plans to create a state shipping
company imply a fundamental political change. In addition, the
principal earner of foreign exchange is now the oil industry, which is

largely under public ownership. The large-scale subsidisation of the rest of Norwegian industry also means that it is more exposed to public demands.

A manufacturing sector that depends on public support to survive must also yield to public demands, and in Norway this situation has given the state direct control of the levels of investment income and profits. The wage and price freeze imposed in September 1978 for a duration of 15 months was aimed at restoring private profits and stimulating private investment. The measures also reflect a degree of desperation, as it was the first time in Norway that the right of unions to negotiate had been suspended.

Even if the measures aiming at restoring private profits and at stimulating the private accumulation of capital should be moderately successful, the historical balance between the private and the public sector is unlikely to be restored. The wages squeeze also has a positive effect on the profits of public enterprises. In addition, because of its dominance of the oil sector, the public sector as a whole has a structural advantage in relation to the private sector. Finally, the channelling of oil revenues to the public treasuries gives the state a structural advantage in relation to private industry as well.

Consequently, the balance of power between the private and public sectors appears to be possibly irreversibly changed. In both the UK and Norway the public sector will have a much improved ability to generate income and to accumulate capital than historically, and compared to most other capitalist countries. This change has profound political implications. To the extent that the public sector cannot generate its own income, for example through public enterprise, income has to be transferred from outside, from the private sector. This transfer of income can take place in two ways: directly through taxes and duties on private business operations and indirectly through taxes and duties on wage-earners' income and consumption. In most advanced countries, not least in the UK and Norway, the residual income requirement of the public sector, i.e. what is not covered by public enterprise, generally represents an important part of the gross national income, and the transfer of income from the private sector mostly takes place in an indirect way, i.e. through taxation of wage-earners' incomes and consumption. In this way the public sector becomes strongly dependent upon the private sector. The reason is that the level of activity of the private sector is decisive for the revenue of the state. The large social budget of the welfare state, as it exists in the UK and Scandinavia, makes the state particularly vulnerable to fluctuations

in its income base. This situation can be described as the dilemma of
the welfare state: on the one hand, the public sector is to provide
services that the private sector supposedly is unable to provide, or from
which it is to be liberated; but on the other hand, the public sector is
unable to finance its activities and its income is a function of the
activities of the private sector, which also creates many of the problems
which are the subject of the welfare state. In this way the large budgets
of the welfare state and its large share of the gross national product do
not isolate the public sector from the demands of the private sector. On
the contrary, the large income requirements of the welfare state make
it particularly exposed to the demands of the private sector.[13] Because
the activities of the private sector secure the income of the welfare
state, a key task for its economic policy is to stimulate the activity and
growth of the private sector.

The dependence is also expressed in the control of the private
sector over the larger part of productive life, employment, investments
and exports. A high level of economic activity generally requires a high
level of investment. Exports are crucial to the maintenance of the trade
balance. When the overwhelming part of productive life is in private
hands, considerations of capital income and private profits must have
a high priority in the formulation and implementation of economic
policy.

This is also true for any leftist governments that want to make the
distribution of income more equal. The reason is that private production, investment and exports depend upon anticipations of private
profits. Any economic policy that to a considerable extent does not
meet these requirements will create serious economic problems for
the welfare state in the form of declining industrial production,
investments and exports, rising unemployment, problems of balance of
payments and a deteriorating income base for the public budgets. In
this way, there is also a political dependence of the welfare state in
relation to the private sector. Normally, the private sector can be
stimulated to produce, export and invest, but it cannot be forced. Thus,
any government that does not want to engage in a trial of strength with
the private sector has to opt for a relationship of confidence and
respect certain minimum private demands concerning capital income
and profits. This is perhaps particularly true of leftist governments.
Through its domination of productive life, the private sector is also able
to exercise considerable control over regional policy. For example,
private investment in industry will normally make the public sector
follow up with investment in services and infrastructure, whereas the

contrary is mostly not the case.[14] The dependence of the welfare state upon the private sector reduces the freedom of action of governments in the choice of macro-economic instruments. In this way, the choice of solutions to economic problems and crises is also restricted. This is the general pattern in advanced capitalist societies, and has a particular relevance for the UK and Scandinavian welfare states.

Through oil and the North Sea model this pattern of dependence is changing, moderately in the UK, drastically in Norway. Oil revenues, through taxation and through direct participation, make the public sector more independent financially. In the UK oil revenues will represent a moderate part of the budget, in Norway they will correspond to a large part of it. The exports of oil from Norway and the substitution of imports in the UK will to a certain extent secure the balance of payments. In addition, as already mentioned, a certain part of the non-oil economy is coming under public control, directly through nationalisation or indirectly through support and subsidies. The political effect of the subsidisation of private enterprise is that it becomes dependent on the state, enabling the state to set standards for its behaviour. As the public sector, essentially through its control of oil, becomes less dependent upon the private sector, to a certain extent the relationship of dependence is reversed. In this way, the UK welfare state to a moderate extent, and the Norwegian one to a large extent, become more isolated from the demands of the private sector. A concrete result is that general considerations of capital income and private profits become less important for economic policy. In this way, the freedom of action of the state in the choice of macro-economic instruments is enlarged.

When private production, exports and investment have a reduced importance in the context of a national economy, the decision processes in the private sector have a reduced importance for economic development and carry less political weight. In this way, the potential of the private sector to defend its own interests is reduced. Correspondingly, the decision processes in the public sector become more important for economic development, and carry more political weight. This raises the question of the distribution of power within the public sector, and this is also a question of the organisation of the state.

The Pattern of Economic Organisation

In the 1970s a new pattern of economic organisation has begun

emerging in the UK and Norway, partly under the impact of oil and
the North Sea model and partly under the impact of the international
recession. The characteristics of the new pattern are, as already pointed
out, increasing state participation in the economy and increased
regulation and control of economic life in general.

If problems of technology and industrial organisation are mastered,
the oil sector will constitute an important instrument for influencing
and regulating the UK and Norwegian economies. The level of GNP
and the gross level of exports will increasingly depend on oil production, particularly in Norway. Thus, provided the governments can
control the level of oil production, it can be used to regulate international accounts. In this respect, the UK and Norwegian governments
will be in a better position than most other advanced industrial
countries, particularly in the control of their exchange rates. Also, in
terms of overall macro-economic management, they will be in a more
favourable position. As the oil sector, and especially the public oil
companies, become a major market for domestic mechanical,
engineering and construction industries, the governments will be able
to influence the level of activity in these branches of industry. This
also provides new opportunities of economic management at the
sectoral and regional level. Through the public oil companies demand
can be channelled quite selectively to desired industries and areas.

Besides intervention through the oil sector, the oil revenues
channelled to the state permit a more active industrial policy. In the
UK, the role which was envisaged for the National Enterprise Board
in implementing industrial policy and in channelling the oil revenues is
an excellent example. Its role in planning and taking entrepreneurial
initiative would not only have increased the size of the public sector
in the economy, but would also have improved the government's
ability to select a particular emphasis in industrial policy.

In Norway, the idea of a public enterprise holding company has
been rejected, and instead oil revenues will be channelled through the
existing state and commercial banks, which will implement the
government's industrial policy. Norwegian industrial policy is likely to
be less detailed and less selectively interventionist than the UK's,
relying instead on general criteria of economic and market performance.
In the event of a prolonged international recession, Norway may,
however, have to opt for a more selective and interventionist industrial
policy in order to maintain employment levels, satisfy regional
considerations and assure technological development. Also in Norway
the use of oil revenues for industrial purposes will strengthen the state

in relation to private industry, and make private industry more dependent upon access to public funds.

As already mentioned, in both the UK and Norway public ownership of industries outside oil has been increasing in the 1970s. In Norway, in the mid-1970s commercial banks have come under closer public control through the 'democratisation' of the banking system, which involves Parliament electing the majority of the board members of commercial banks. The non-socialist opposition has accurately described this as nationalisation in disguise. This trend is likely to continue. In Norway, nationalisation of the major shipyards would conform to UK and Swedish policies. In the UK, control of the oil sector could perhaps lead to heavier public involvement in the chemical industry.

The combined impact of these measures is a new pattern of economic organisation in the two countries. The state now has a much more important role in economic life than ever before, and its role exceeds that of government in most other capitalist countries. The term state capitalism seems quite appropriate. The UK and Norwegian economies are still essentially market economies, but the public sector increasingly dominates these markets.

In the present UK and Norwegian economies resources are allocated both by the market mechanism and by the political and administrative process. The roles and limits of the two methods of resource allocation have never been clearly defined, so that the high level of public intervention in the market that currently characterises the two economies is fairly haphazard and poorly co-ordinated. This creates confusion and inadequate control of public expenditure on industrial and regional policy. In fact public intervention has so far been mostly defensive, aiming at saving employment and technology rather than restructuring the economy.[15] The abundance of public oil revenues creates a risk that this practice may be extended. The challenge is to use the public oil revenues in a more constructive way, aiming at systematic and co-ordinated public intervention in the market with clearly defined goals. This requires not only a definite industrial strategy, but also a redefinition of the role of the state in the economy. Specifically, this means a new definition of the relationship between the state and private capital.

In any case, a new form of economic system is emerging in the UK and Norway, and this process is accelerated by the North Sea model, with administrative allocation of resources gradually becoming more important in relation to allocation of resources through the marketplace. This is also a transfer of power from private industrial and financial capitalists to politicians and administrators in the public sector, so that gradually a new power structure is emerging as well in the two countries.

Oil – Capital, Labour and the State

According to orthodox Marxist theory, the essential function of the bourgeois state is to secure the conditions of reproduction of private capital. In the mid-1970s the UK and Norwegian states have had great difficulty in fulfilling this role. The UK state has been unable to prevent wage increases, which critically eroded private profits. The Norwegian state, through its economic policy, has in this particular perspective acted in an eminently revolutionary way, albeit fairly unintentionally. The wages squeezes in the UK in 1976 and in Norway in 1979 may appear acceptable as one-off measures, but they are politically extremely difficult to repeat and they are likely to produce accumulated wage demands. Thus, the problem of the profits squeeze in the private sector is likely to appear again, perhaps as a permanent phenomenon, unless there are much higher rates of unemployment or an authoritarian control of trade unions, both of which appear politically difficult as well. Against this background, UK and Norwegian capitalism, and the UK and Norwegian states, seem to be stuck in a tricky situation, and the two societies in many ways appear to be at an important historical crossroads, where the extreme alternatives are a more authoritarian form of capitalism or a further socialisation of the economy. This gives potential for an increasing level of conflict in the two societies, and eventually for political destabilisation. The crucial question is whether oil in this situation can provide a third way out, with improving prospects for both capital and labour, and with a politically stabilising effect.

In theory, oil can provide an alternative solution. Oil revenues and their impact on the balance of payments enable a higher growth of production and consumption than otherwise would have been possible in the UK and Norway. Consequently, oil gives more room for expanding real wages and private profits, and gives a certain amount of room for expanding both at the same time. Thus, oil can also be politically stabilising. In this perspective neither a repeated wages squeeze nor a renewed profits squeeze need be necessary, and higher rates of unemployment or an authoritarian incomes policy may be

avoided as well. This is a possible outcome, but it requires a fine balance
of the correct macro-economic measures, an increased propensity of
capitalists to invest, and trade union restraint.

In practice, oil can affect the distribution of income and stimulate
increasing wage demands. The private propensity to invest not only
depends upon capital being available at acceptable rates of interest, for
example through available public oil revenues, but also on anticipated
profits. In the contexts of the UK and Norway, with effective trade
unions, there is a fairly strong urge for the equalisation of incomes and
living standards. Therefore, increasing private profits, for example as a
result of domestic use of oil revenues, have a fair chance of stimulating
wage claims that again erode profits and negatively affect investment.
This process may be modified by an extension of industrial democracy
that is gradually taking place in the UK and Norway, giving labour a
more direct stake in management and a more direct interest in the
long-term situation of enterprises. But if this is to be effectively a
moderator on wage claims, with positive effects for investment, labour
participation in management must have a direct influence on the
distribution and use of profits. This again implies a lower share of the
profits going to the owners, the capitalists, with negative effects for
the capital market and the mobility of capital, but with potentially
positive effects for the self-financing of enterprises. Otherwise, a large-
scale channelling of public oil revenues to private industry may lead
to increasing real wages rather than to increasing investment. In any
case, in order to be politically acceptable, any increase in the level of
private profits must be accompanied by a certain degree of real wages
increase. Otherwise, a significant increase in private profits could
easily provoke substantial and fairly irresistible wage claims. Against
this background, a restoration of capital income and private profits
to their previous levels seems politically difficult in the UK and
Norway, and more difficult than in other capitalist countries that do
not have oil.

In this situation, capital income and private profits may appear as
an unnecessary and perhaps risky link in the process of capital
accumulation. In both countries, the Labour governments envisaged
channelling investment funds, financed by oil revenues, directly
through public agencies, short-circuiting the private capital market.
This is also a substitution of the market mechanism by administrative
allocation, and substituting private owners of capital by politicians
and top civil servants as the controllers of the process of accumulation.
This is a gradual change, but nevertheless of great importance, as it

separates elements of dynamism in the two economies from private
capital. Against this background, oil could prolong and even intensify
the chronic crisis of UK capitalism. In Norway, oil gave the basis for an
economic policy that contributed to create an acute crisis for
Norwegian capitalism, and this crisis may well become fairly chronic.
Such prospects should worry the owners of private capital, and indeed
there is already a fair amount of frustration among the core of capital
owners in both countries that the North Sea model to a large extent
neutralises the oil revenues from their point of view, and even reduces
their own importance in the national economies. In the UK, private
control of the oil revenues would have had a highly stimulating effect
on the capital market and the City of London. In Norway, a largely
private control of the oil industry would have given shipowners a
historic chance of pursuing expansion on a more stable basis and to
offset the decline in shipping profits. Consequently, there are strong
interest groups in both countries that argue for a considerable
modification, or even a dismantling, of the North Sea model.

For the new Conservative government in the UK and potentially
for a non-socialist government in Norway, it would be tempting at
least to reduce the degree of state participation in order to give private
industry a higher stake in the oil industry, and in order to channel a
larger proportion of the oil revenue to the private sector. But such a
measure would also have negative effects upon public revenue, reducing
freedom of action in economic policy. Also, there is the risk that
increasing private revenues from oil might trigger off high wage
claims, commented upon earlier. Thus, there is quite a chance that
the North Sea model would not be substantially altered in either
country by conservative governments. The Conservative government
in the UK has in the summer and autumn of 1979 largely retained the
oil policies inherited from the Labour government. Consequently, in
this situation any UK or Norwegian government seems likely to
preside over the gradual socialisation of the economy, in some cases
without great enthusiasm, that results from the dynamics of the North
Sea model, combined with a depressed international economic context.

The decline of private capital in the two countries corresponds to a
growth in the power of the public sector. This is a highly ambiguous
phenomenon. First of all, there is the expansion of public enterprises,
primarily in the oil industry but also in some other sectors. The
expansion of capital-intensive, profitable public enterprises implies
that a part of the economic surplus is withheld from private owners of
capital, eroding the principle of capitalism that the surplus essentially

is to be controlled by private capitalists. In this way, the growth of
public enterprise implies a break with the traditional form of
capitalism. On the other hand, the public enterprises are generally run
according to business criteria, emphasising micro-economic profitability, and they operate in a market. In this way, the growth of public
enterprise represents by itself no break with the market economy, nor
with the principle of accumulating capital at the micro-economic level.
Therefore, state capitalism is not synonymous with socialism in an
orthodox Marxist sense, implying the preponderance of national
economic criteria and capital accumulation at the level of the state. It
can thus be argued that the growth of public enterprise is rather a
restructuring of capitalism, according to new needs and premisses.[16]
But on the other hand, the growth of public enterprise also reduces the
importance of the private *rentier*, and thus the incentive to produce
according to the needs of private capital income.

Second, there is the expansion of public oil revenues, which
represents a small part of the gross national product in the UK, a
larger part in Norway. The expansion of public revenues from profitable
business operations implies that part of the economic surplus is
withheld from the participating enterprises, eroding the principle of
capitalism that the surplus is essentially to be controlled at the micro-
economic level. In this way, the growth of public oil revenues implies
a break with capitalist practices. In addition, the largely administrative
allocation of these revenues, where political considerations may be as
important as economic ones, also implies a break with the market
economy. Thus, the centralisation of oil revenues with the Treasuries
apparently is a non-capitalist practice, and perhaps even fairly socialistic
in an orthodox Marxist sense.

The rising public sector comprises both expansive public firms and
an expansive state. The relationship between the public enterprises and
the state will be of critical importance for the political economy of the
UK and Norway in the coming years. A possible outcome is that
private capital increasingly will be substituted by profitable, capital-
intensive public enterprises in key sectors of the economy which will
be able to exercise a greater political influence than private firms, and
to influence public policy more strongly. Another possible outcome
is that gradually, through the increasing dominance of the state in the
process of capital accumulation, the economy will function more
according to social needs and to social criteria of efficiency.

When the state owns enterprises that are active in a market
economy, the state becomes a party in the market and acquires

commercial interests in addition to its administrative interests. In this way, the traditional controlling function of the state in relation to the market economy is supplemented by a participatory function. This participatory function can reinforce the controlling function, as it gives the state more instruments to influence economic life, but this implies that the central economic organs of the state have a sufficient degree of control over the entire public sector in the economy, in particular the public enterprises. If this is not the case, the public enterprises may establish themselves as centres of power, and possibly dominate their subordinate administrative organs. In this way, there may be a conflict between the participatory function and the controlling function, and the state can run into conflicts with itself. The outcome can be a struggle of competence between the central economic organs of the state and the sectoral administrative organs, with negative effects for the co-ordination and effectiveness of public policy. This is also a question of how the political process is activated in relation to the public sector of the economy. Insufficient political attention can lead to inadequate parliamentary control over the administration and to a weakening of the central economic administration. On the other hand, a high level of political attention may lead to a more efficient parliamentary control over the administration, and to a strengthening of the central economic administration.

Against this background, there is an evident possibility that the growth of the public sector in the UK and Norway, linked to oil development, will essentially mean the growth in power and influence of large public corporations. These may be more difficult to control from outside than large private corporations.[17] They may also have a comparative advantage in their close relations with the public administration, enabling them better to influence administrative decisions, and in particular the administrative organs that are responsible for the corresponding sector. Thus, there may be a co-ordination of the administrative and the commercial interests of the state on the premises of the latter, i.e. a rising corporatism.[18] In this way, the rise of the public sector may also imply a transfer of power from the state itself to the large state enterprises. Potential candidates are evidently the national oil companies in both countries, in the UK the National Enterprise Board and its subsidiaries, in Norway the large public corporations. This also means a fragmentation of the state and increasing problems of implementing a comprehensive economic policy. Such prospects are to a certain extent offset by the state directly becoming an accumulator of capital, but the rise of the public enterprises essentially creates a need for more structured economic planning in order to maintain the control of the state, with a clear definition of objectives and a more detailed implementation. Otherwise, the public enterprises are likely to behave increasingly like private firms, and through increasing links with private interests, the rise of public enterprises could ultimately lead to a reprivatisation of the economies of the UK and Norway.

Notes

1. Øystein Noreng, 'Oljen og Norges politiske økonomi', *Kontrast*, no. 3/4 (1974), pp. 11–24.

2. Øystein Noreng, 'Staten, oljen og de politiske kanaler' in *Om staten* (Pax Forlag, Oslo, 1978), pp. 136–76.

3. *Politische Ökonomie des heutigen Monopolkapitalismus* (Dietz Verlag, Berlin (East), 1972), pp. 384 ff.

4. *New York Times*, 18 June 1978.

5. Andrew Glyn and Bob Sutcliffe, *British Capitalism, Workers and the Profit Squeeze* (Penguin Books, Harmondsworth, 1972), pp. 15 ff.

6. Stuart Holland, *The Socialist Challenge* (Quartet Books, London, 1975), p. 56.

7. Glyn and Sutcliffe, *British Capitalism*, pp. 15 ff.

8. Holland, *The Socialist Challenge*, pp. 394 f.

9. Ibid., pp. 395 f.

10. *The Economist*, 9 September 1978, p. 15.

11. *Revidert nasjonalbudsjett 1978*, St.meld.nr. 82 (1977–8), (Ministry of Finance, Oslo, 1978), p. 46.

12. Trygve Thu, 'Klar katastrofealarm for norsk industri', *Teknisk Ukeblad*, 17 August 1978, pp. 12–13.

13. S. M. Miller, 'Planning: Can it Make a Difference in Capitalist America?', *Social Policy*, no. 2 (1975), pp. 12–22.

14. Holland, *The Socialist Challenge*, pp. 121 ff.

15. Ibid., pp. 120 ff.

16. Alain Touraine, *La société post-industrielle* (Editions Denoël, Paris, 1968), pp. 68 ff.

17. Romano Prodi, 'Italy', in Raymond Vernon (ed.), *Big Business and the State* (Harvard University Press, Cambridge, Mass., 1974), pp. 45–63.

18. Øyvind Østerud, *Samfunnsplanlegging og politisk system* (Gyldendal, Oslo, 1972), pp. 59 ff.

9 INTERNATIONAL DIMENSIONS

North Sea Oil and the Western World

In quantitative terms the oil and gas from the UK and Norway make
only a modest contribution to the energy self-sufficiency of Western
Europe, and the contribution to the OECD area seen as a whole is
fairly minimal. By 1985, the petroleum production in the UK and
Norwegian parts of the North Sea could be 250 mtoe, corresponding
to less than 15 per cent of the anticipated energy demand of Western
Europe (1,700 mtoe).[1] It would supply only about 21 per cent of the
oil and gas needs, so that Western Europe as a region will still be heavily
dependent on oil and gas imports. For the entire OECD area, UK and
Norwegian oil and gas production will correspond to only 5 per cent
of the total energy demand by 1985 (5,100 mtoe).[2]

There are, however, qualitative and political aspects of North Sea
oil and gas that make them very important for the Western world. In
addition to being located close to major markets, the sources of oil and
gas are located in industrial countries that are traditionally part of the
Western world, and which are linked to the consumer countries through
extensive trade, finance and even co-operation in the field of military
and security policy. This implies a fairly undisputed security of supply,
causing the political rent mentioned previously, which means that if all
other factors are equal, North Sea oil and gas are likely to be more
in demand than corresponding types of oil or gas from other suppliers.
Furthermore, any supplies of oil and gas from within the OECD area
can be seen as contributions to reducing, or moderating, a dependence
upon external sources of petroleum judged economically unhealthy and
politically potentially dangerous. Consequently, for reasons of energy
policy, economic policy and foreign policy, supplies of North Sea oil
and gas can be seen as more valuable than corresponding supplies from
outside the OECD area. Therefore, there can also be a qualitatively
different need for North Sea oil and gas than for oil and gas imports,
even if external supplies should appear as momentarily unlimited. Thus,
demand for North Sea oil and gas can be more active than demand for
external supplies of petroleum. This active demand can also take the
form of outright pressure, depending on the outlook for world
petroleum. Consequently, the counterpart of the political rent for

North Sea oil and gas, which may be a commercial advantage, is that the suppliers of North Sea oil and gas are potentially subject to stronger pressures than external suppliers, and this may be a serious political disadvantage.

The major flaw of the North Sea model, as it was basically conceived initially in Norway and subsequently adopted in the UK, is that it treats the international aspect of the problem lightly, if at all. This is a more serious problem for Norway than for the UK, because of differences in potential petroleum exports and differences in size. The North Sea model, implying a good deal of 'oil nationalism' and emphasising that the level of development be set essentially according to national needs and state control as well as the preference for national companies, could reasonably appear as fairly provocative to the major allies and trade partners of the UK and Norway, particularly in the aftermath of 1973. Instead of actively contributing to offset the damage, real or supposed, done to the Western economy by the 'oil revolution', the British and the Norwegians could be seen as actively following up the work done by Arabs and other OPEC countries, and consequently as legitimating both the price rise and the oil nationalisation, and for reasons of pure self-interest abandoning any solidarity with the rest of the West. This has a particular relevance for Norway, whose 1974 oil plan, emphasising a moderate rate of development plus full national control, was published in the immediate aftermath of the oil crisis. For any observer not particularly familiar with the local Norwegian scene, the new policy could at best appear as excessively simplistic, or even provincial, and at worst as cynically egotistical. The optimism expressed concerning the rest of the Norwegian economy, given the high dependence on exports and international economic conditions, could appear as simplistic or even naïve, revealing a lack of understanding for the problems of Western countries that do not have oil to export.

The production policy, keeping the level of output down at least implicitly as a function of higher prices, and setting the level of production by purely domestic needs for income, could be seen as narrowly egotistic, or perhaps even as cynically withholding supplies in order to have a permanently tight world oil market, contributing to future price rises. Such points of view may not fully explain the motivations behind Norwegian oil policy, but they do concern a country which is highly dependent upon a diversified foreign trade with non-oil exporting industrial countries, and which wants to maintain a diversified pattern of exports. On the other hand, there was undeniably in Norway and in the Norwegian Labour Party a certain respect and even enthusiasm for the 'oil revolution', and as already pointed out, it did have an impact. In addition, there was in the UK and in the British Labour Party an explicit respect for Norwegian oil policy. Thus, OPEC directly influenced Norwegian oil policy, and at least indirectly UK oil policy. This may have contributed to considerations for oil consumers being largely ignored.

In any case, international considerations should be no major problem for the UK, as the potential for net exports of oil is fairly limited over time, even with a rapid depletion policy. In addition, the UK is in the European context a major power and not easily subject to foreign pressure on her domestic policies. Thus, within the EEC, the OECD or the IEA, the potential gain from putting pressure on the UK to increase the rate of production appears fairly small, and the chances of such pressures succeeding also appear limited. With Norway, the situation is the opposite.

Given low domestic consumption, Norway appears as a potential important net exporter of petroleum, with exports depending directly upon the level of development. Even with her oil wealth, Norway is a minor European country, extremely dependent upon others; Norway cannot even feed herself, giving an apparent exposure to foreign pressures. The potential gain from putting pressures on Norwegian oil policy may appear considerable. For example, in a situation of scarcity, other countries could link deliveries of Norwegian oil and gas to the purchase of goods and services, or to deliveries to Norway. In a more normal situation, the pressures are likely to take more subtle forms. For example, the coupling of oil to industrial co-operation could be such an element. On the one hand, it may be wise to take out the political rent in the form of access to industrial experience and technology, and to compensate the depletion of a finite resource with access to more durable industrial structures. On the other hand, this can reflect a lack of comprehensive industrial policy, with foreign interests providing jobs and technology, but also compromising the moderate production policy and national control of the oil industry.

> The North Sea model has not been designed for the interests of oil importing OECD countries, and these countries have an interest that production targets be revised upwards in the North Sea, and that foreign interests be given a larger share.[3]

An accelerated development of UK and Norwegian oil can have a

Table 9.1: North Sea Oil and Western Europe 1975–2000 (mtoe)

		1975	1980	1985	1990	1995	2000
Energy	(a)	1,190	1,346	1,523	1,723	1,950	2,207
demand	(b)	1,190	1,413	1,679	1,994	2,368	2,812
Coal		245	275	290	320	373	435
Hydro		83	86	92	100	105	110
Nuclear		22	68	137	258	330	420
Gas		148	213	253	281	356	450
Oil demand	(a)	682	704	751	764	786	792
	(b)	682	771	907	935	1,204	1,397
North Sea	(c)	9	130	170	135	85	55
oil	(d)	9	130	170	195	195	150
	(e)	9	130	170	225	225	180
Oil imports	(a) (c)	673	574	581	629	701	737
	(a) (d)	673	574	581	569	591	642
	(a) (e)	673	574	581	539	531	612
	(b) (c)	673	574	581	710	1,119	1,217
	(b) (d)	673	574	581	650	1,009	1,122
	(b) (e)	673	574	581	620	979	1,092

Assumptions:
(a) Energy demand growing at 2.5 per cent a year.
(b) Energy demand growing at 3.5 per cent a year.
(c) British oil production peaking in 1985, Norwegian oil production at low
    scenario.
(d) British oil production keeping ceiling 1985–95, Norwegian production at
    medium scenario.
(e) British oil production keeping ceiling 1985–95, Norwegian production at
    high scenario.

considerable impact upon Western Europe's self-sufficiency in energy, and consequently upon the freedom of action in economic policy in relation to a given target for oil imports. Thus, the UK's and Norway's trade partners have a legitimate interest in a high level of production, and they could even argue that this will positively affect the rest of the UK and Norwegian economies, as Western Europe then can assume a higher rate of economic growth, with positive effects on demand for other UK and Norwegian goods and services. Particularly, if the world's energy problem is seen as that of muddling through the rest of this century on the basis of oil, awaiting a breakthrough of alternatives such as nuclear fusion and solar energy, there is a fair argument that a high rate of development for UK and Norwegian oil can be of a certain assistance during the transition period, helping to maintain acceptable economic growth rates.

For the UK and Norway things are not as simple, even if the argument that their oil policy does have an international impact is accepted. For the UK the problem is that a high rate of production in the mid-1980s may perhaps moderately contribute to postpone scarcity and a new price rise, but then scarcity and the price rise could occur as UK oil and gas production declines, leaving the UK to import increasing quantities of increasingly expensive oil towards the end of this century. It is in the interest of the UK to have an opposite price scenario, with the price rise occurring as UK oil production is building up, reinforcing its competitive advantage, and with oil prices falling again as oil production declines because of depleted reserves. For Norway the problem is more complicated. With substantial new discoveries, Norway will be able to maintain and possibly substantially expand production. In this way Norwegian oil could become gradually more important, for Western Europe and in the world market, with Norwegian policy perhaps becoming more relevant for the price development. In such a case Norway could, at a lesser scale, be facing some of the same problems of choice that Saudi Arabia does. On the one hand, Norway's interests would be to protect other economic interests, i.e. essentially non-oil exports that are employment-intensive, and eventually to avoid a rapid price rise for oil. This could imply an increasing production to moderate oil prices. On the other hand, Norway would also have an interest in protecting her oil reserves from being depleted at a rapid rate, and also in avoiding overheating the domestic economy. This could imply a ceiling on oil production. Internationally, there is likely to be scarce sympathy for such dilemmas, unless the world's energy outlook changes drastically, in which case the dilemma would cease to be.

The Question of Siding

Given interests in oil production and oil prices which diverge radically from those of practically all other Western countries, and which in many ways are close to those of the OPEC countries, the situation of being members of the Western industrial world and its institutions is not without problems for the UK and Norway. In relation to the world oil market, both countries have a vested interest that OPEC does not break down, with a collapse of the oil price. It is equally important for both countries that the OECD economies do not suffer an economic setback because of a scarcity of oil, or because of a sudden jump in the price of oil. This implies that the relationship between oil policy, economic policy and foreign policy be reviewed carefully and continuously.

The problem of siding presented itself in the immediate aftermath of the oil crisis. Both the UK and Norway were invited to the Washington meeting in February 1974, where Dr Kissinger presided over the creation of the Energy Coordination Group, ECG. The ECG was widely, and probably rightly, seen as an attempt to make a counter-cartel to OPEC in order to defend the interests of the consuming countries in matters of production and prices. For the UK, given the limited potential for increasing oil exports, and being a major European power, membership was fairly unproblematic. In Norway, however, the affiliation with the ECG created tensions in the administration and in the Labour Party, because of fear of being subject to strong foreign pressure. Indeed, to join an oil consumers' cartel a few months after publishing the 1974 oil plan could be seen as a sign of political schizophrenia within the Norwegian government.

In this document, Norway's community of interests with the OPEC countries in matters of prices and production had been explicitly stated, as well as the need to develop closer contacts with the OPEC countries.[4] In 1974 there was increasing contact between Norway and OPEC itself, partly through the Norwegian embassy and OPEC head-quarters in Vienna, and in the autumn of 1974 OPEC's Secretary General, Abderrahman Khene, paid an official visit to Norway. At the same time, in 1974, the ECG was transformed into a more permanent organisation, the International Energy Agency, IEA. The UK joined without hesitation. In Norway, the IEA issue became a matter of intensive strife within both the administration and the Labour Party.[5] It should be recalled that two years before, in 1972, a referendum had rejected EEC membership. The opposition to EEC membership had consisted of farmers, fishermen, smaller businessmen, the bulk of the trade unions, leftist intellectuals, etc. It was largely financed by the farmers' co-operatives, but it gave leftist intellectuals a unique platform. At the elections of 1973 the Labour Party was roundly defeated, and got into power only because of a substantial left socialist representation. In this political climate it could be feared that the anti-EEC alliance could be reactivated for purposes of oil and the IEA, with substantial public support, perhaps even compromising Norway's links with the Western world. In addition, there was in the Norwegian administration and in the Labour Party considerable sympathy for OPEC, which was seen by many as a champion for just claims from the Third World, and there was a corresponding lack of enthusiasm for the United States government and Dr Kissinger, partly on the background of the continued war in Vietnam; Watergate also had an impact. Against this background, there was considerable opposition to Norway joining the IEA, in spite of persistent pressure from the United States. In the late autumn of 1974 Norway decided not to become a full member of the IEA, but to obtain an association agreement. In the spring of 1975 Norway got an associated status with the IEA, participating in all essential activities, but not being part of the crisis planning. New problems in the relationship with the IEA arose in 1976, with the long-term programme, which envisaged non-discriminatory practices and freedom of investment in energy development of other member countries, as well as a guaranteed floor price. Norway finally, again after pressures from the United States government, ratified the long-term programme, but with reservations concerning the non-discriminatory practices and the freedom of investment. The IEA agreement aroused some political opposition at the outset, but this has largely subsided.

In recent years, there has been a certain attempt at balancing the IEA association with closer OPEC links. There have been rumours that Norway wanted to join OPEC.[6] There were also rumours that a Norwegian application had been rejected on the grounds that it was not a developing country. In any case, there have been increasing contacts, with official visits in both directions. Contacts are becoming more regular and more structured, but they do not yet take the form of regular consultations. Norway is building up contacts with OAPEC. An example is a joint conference held in Oslo in September 1978. At the industrial level, links are increasing fairly rapidly between Norwegian industry and OAPEC countries.

For the UK oil does not present any serious problem of foreign policy, for reasons mentioned already. For Norway oil does potentially present serious problems of foreign policy, at two different levels. First of all, any Norwegian government in the question of depletion policy and partly in the question of organisational pattern and state oil control must make a trade-off between domestic and international considerations. As pointed out, there is a wide consensus in Norway on the principles of a moderate rate of extraction and a high degree of state control. Some people in Norway fear foreign pressure imposing a higher rate of production and lower degree of state control. In some quarters this fear takes the character of an obsession, even if recent economic problems may have contributed to greater tolerance of a higher rate of development. Consequently, many Norwegian politicians and civil servants have a fear of provoking an unpredictable public reaction through changes in oil policy. But they also fear foreign pressures aiming at a higher rate of production. Second, Norway has structurally split interests in the world oil market. There is a paradox that, by following what is often seen as a defence of the oil interests, i.e. by keeping a low rate of production, the oil interests do not become very important. Conversely, there is an equal paradox that, by yielding to foreign pressures, by keeping a high rate of production, Norway's oil interests become more important, and so does the community of interests with OPEC.

Presently, Norway has ambiguous relationships with both OPEC and the IEA. This partly reflects an astute strategy, partly a profound dilemma. The relative glut of oil in the world market and the easing of tensions between OPEC and the IEA have improved Norway's freedom of action. With increasing scarcity of oil, as predicted by many for the late 1980s, rising prices, and perhaps a renewed tension between OPEC and the IEA, Norway's freedom of action will be reduced again, nationally and internationally. In case of an acute oil crisis, Norway might not be able to contribute very much, even if she wanted to. The problem is rather that there may be a more chronic, or creeping, oil scarcity building up, and in such a situation additional supplies from a number of producers would help, and also from Norway. So far, there has in OPEC and OAPEC circles been a considerable sympathy for Norwegian policy.[7] In a situation with a gradual scarcity on the horizon, any unwillingness on the part of Norway to increase the rate of production of potentially large reserves could easily be seen as creating an additional pressure on the traditional producers, and consequently as a lack of solidarity in this respect as well. Thus, the question for Norway might not any longer be to take sides internationally, but rather to choose between international and domestic considerations. In the present economic situation, a gradual increase of the level of activity would probably be politically tolerable, and even perhaps be seen as desirable. But this does not reduce the need to devise a far-sighted political strategy.

Given the strategic and economic importance of oil, particularly when scarcity looms on the horizon, exporting oil is also a form of foreign policy. Because of small net exports, this has little relevance for the UK. For Norway it is highly relevant. The patterns of marketing Norwegian oil will also create links of economic and political dependence. This raises the question of the relationship between commercial and political considerations in marketing petroleum. The easy way out is to consider commercial aspects only, but the political aspects cannot be defined away. For example, commercially speaking, it could be most interesting to sell a large part of Norwegian oil in the world's leading market for low-sulphur oil, the East Coast of the United States, where demand is high and the ability to pay beyond doubt. Politically, such a solution might alienate Norway's immediate neighbours in Scandinavia and the rest of Western Europe. Even if deliveries of North Sea oil to the United States are offset by reduced United States demand for Libyan and Nigerian oil, for example, there is still a perceived difference in security of supply.

Against this background a marketing strategy is now being conceived where Norwegian oil is part of wider political agreements concerning industrial co-operation and trade. A comprehensive agreement with Sweden was prepared but was turned down by the Volvo shareholders.

Scandinavian co-operation might benefit from Norwegian oil.[8] Sweden and Denmark are among those countries of Europe that are most dependent upon oil imports, and these markets are geographically close. In addition, the cultural affinity makes industrial and trade co-operation feasible, particularly with Sweden. This could also be politically wise. In Europe, as in the world, there is a mismatch between populations and natural resources.[9] Norway belongs to an outer fringe of OECD countries, where the imbalance between the population base and the resource base has the opposite sign from most other countries. This is economically an asset, but politically a potential liability, particularly when the population base is only 4 million. By actively promoting Scandinavian co-operation, and giving other Scandinavians a certain preference in exploiting Norway's natural resources, the balance between population base and resource base can be somewhat improved. Economically, this can give a more rational exploitation of the resource and greater spin-off. Politically, this can provide alliances and support that could be of crucial importance in a situation with less freedom of action. As the development moves further north, the question of an enlarged demographic, economic and political base could become more relevant.

The Northern Waters

On 1 January 1977 Norway extended her sovereignty over an economic zone stretching 200 nautical miles from the coast, or until the median line with other countries' economic zones. The motivation was to control an area that has an increasing economic interest, and in which any activity directly concerns Norwegian fishing, petroleum and security interests. The move had been prepared for several years, but the direct precedents were the decisions taken by the United States, Canada, Mexico and India in 1976 to extend their economic zones to 200 nautical miles.[10] The move had a dramatic effect upon the size of Norwegian territory. As already mentioned, Norwegian territory in the North Sea south of 62° N has an estimated size of *c.* 125,000 square kilometres. The maritime territory north of 62° N, off central and northern Norway, stretching out to the point where the continental shelf falls abruptly towards the ocean floor, making an effective limit to economic activities with present technology, is approximately 875,000 square kilometres, giving Norway a maritime territory of *c.* 1 million square kilometres, two and a half times the size of Norway itself, including Spitzbergen (390,000 square kilometres). The continental shelf of the Spitzbergen archipelago, north of 74° N, is *c.* 350,000 square kilometres. This area was partly considered Norwegian before the extension of the limit. With the economic zone, Norway's jurisdiction is extended over an area of about 1,720,000 square kilometres.[11] In addition, an economic zone around the Norwegian island of Jan Mayen could eventually make up about 350,000 square kilometres, giving a total maritime territory of about 2 million square kilometres.

Thus, from the point of view of total territory, Norwegian jurisdiction could extend over 2.5 million square kilometres, in Europe next only to the Soviet Union. Most of this is in Arctic waters. From the point of view of population, Norway disposes of one of the world's largest territories. This is also an area with an interesting resource potential, carrying an increasing economic interest. It constitutes a base for sustained growth of the Norwegian economy, but it also underlines the reversed mismatch between population and resources in Norway, compared to the rest of Europe. In addition, important questions of international law are yet to be settled.

As already mentioned, along the north Norwegian coast there appears to be a sediment of Tertiary age of promising proportions that could contain significant quantities of oil and gas. Since 1969 seismic surveys have taken place on behalf of the Norwegian government. Along the entire coast there is a sedimentary layer in the form of a wedge, increasing in thickness towards the west. Off central Norway, between Møre and Lofoten, it appears to be a continuation of the North Sea basin, with a thickness of around 3 kilometres off Møre and of 1.5–2 kilometres off Nordland.[12] Further north, off Troms and Finnmark, the geology is apparently different, with age increasing from west to east. These sediments appear to be part of a larger sedimentary basin covering the Barents Sea and the waters around Spitzbergen up to the North Pole, and probably stretching east past Novaya Zemlya into the Kara Sea. This continental shelf of the Arctic Ocean appears fairly promising from a petroleum point of view.[13] However, so far no exploratory drilling has taken place. If the sediments do contain petroleum, the size of the area could make it into one of the world's more important petroleum provinces at a later stage.

For physical and political reasons, the northern waters make up two different areas. The continental shelf off the coast of central and northern Norway, up to Lofoten at approximately 68° N, does not present very extraordinary problems compared to the North Sea. Weather conditions may be worse, with the problem of icing of platforms and equipment in the winter. But this may be offset by generally less depths and shorter distances from land. As the Norwegian continental shelf here ends in the North Atlantic with no immediate neighbours, there should be no political problems. However, a problem might arise if petroleum development is possible outside the 200-mile limit, and if international law should not permit Norway to extend her sovereignty further than this limit, covering areas where economic activities are technically possible. This could, for example, affect part of the Vøring plateau off northern Norway.

The continental shelf off the northernmost part of Norway, off Troms and Finnmark, commencing at approximately 70° N, and stretching up to and past Spitzbergen, presents a series of particular problems. Here the continental shelf meets with the ocean floor at a maximum depth of about 400–500 metres, and often at smaller depths. These areas appear to be particularly interesting from the point of view of petroleum geology.[14] Again, weather conditions are likely to be tough, and the problem of icing more serious, but some of the areas of immediate interest are located at moderate depths not very far from the coast. But a more extensive petroleum development of this area will probably require technological breakthroughs. There are also political problems relating to the delimitation of Norwegian and Soviet waters, Soviet security interests, and the position of Spitzbergen. As for the delimitation, the Norwegian point of view is that the Norwegian continental shelf stretches from the coast of Norway up to the North Pole, with the archipelago of Spitzbergen having no continental shelf of its own, and with the median line constituting the borderline at sea with the continental shelves of other countries, following the precedents from the North Sea.

This interpretation, which is opposed by the Soviet Union, gives Norway full sovereignty and the sole right to control petroleum activities in a vast area stretching from her coast up to the North Pole and well into the Barents Sea. The Soviets maintain that the Spitzbergen archipelago has a continental shelf of its own, which gives all the signatories of the Spitzbergen Treaty equal rights to pursue economic activities there. In addition, the Soviet Union sees the borderline between the continental shelves of the two countries as the sectoral line, a straight line drawn from the North Pole to the point on the mainland where the two countries meet. This principle is contrary to the precedent from the North Sea and most other parts of the world. As a consequence, there is disputed territory between the median line and the sectoral line.

Soviet security interests in the Barents Sea are determined by naval activities in the Murmansk area.[15] The Barents Sea is used as a passage for Soviet warships and submarines, and is potentially of great importance in wartime.[16] In addition, a large part of the Soviet navy is attached to the Murmansk area, and thus the Soviet Union has vital security interests in the Barents Sea. A large number of foreign oil installations could be seen as impeding the passage of ships and as contrary to these vital interests. The potential use of oil installations as sonar detectors of submarine noise could be seen as particularly harmful. However, sonar units can be placed on the bottom of the ocean without oil rigs or platforms.

The Spitzbergen Treaty of 1925 gave Norway sovereignty over the islands, but all partners to the treaty have equal rights to pursue economic activities there. For many years the islands were neglected by Norway, with Norwegian coal production and employment decreasing. The Soviet Union, by contrast, maintains a relatively large group of people on the islands. In recent years Norwegian activities have been increasing, and Norway is more actively asserting its sovereignty over the islands. This is a permanent source of friction between Norway and the Soviet Union. Oil exploration and production in the Barents Sea and the increasing military importance of the area will surely add to this friction.

None of Norway's major allies, in particular the United States and the UK, has endorsed the Norwegian view that the continental shelf around the archipelago is Norwegian without restrictions. These allies could have a strategic interest in comprehensive Norwegian control of the continental shelf up to the North Pole, but given the prospects of petroleum, they have an economic interest in the opposite direction. There is evidently a British and American fear that, if this area were under exclusive Norwegian control, the level of oil activity and the participation of foreign interests would be restricted. The assumption is evidently that a continental shelf governed by the same economic regime as Spitzbergen might produce more oil and might allow more foreign participation.

For the Soviet Union, the prospect of a high level of petroleum activity on the continental shelf around Spitzbergen, and of a high level of foreign participation through large Western oil companies, is clearly not in its interest. Thus, Norwegian sovereignty over the continental shelf around Spitzbergen seems more in line with Soviet interests.[17] The present Soviet stand on this issue appears to be tactically aimed at getting other concessions from Norway. Possible concessions involve the delimitation of the Barents Sea, and a possible Soviet interest in some kind of a bilateral Soviet-Norwegian regime for Spitzbergen. Such Soviet-Norwegian bilateralism is obviously not in the interest of Norway's Western allies. Thus the Western stand could also be tactically motivated and designed to get concessions from Norway. The concessions that would interest the Western allies would be the level of oil development and possibly the degree of participation by foreign companies. As a result of these differences it could ultimately be in the interest of both the Soviet Union and Norway's allies to give her the full jurisdiction of the continental shelf up to the North Pole as a second-best solution. But in order to achieve this, Norway might have to give some concessions to the Soviet Union concerning foreign participation and military installations in certain waters, and some concessions to the Western allies concerning the rate of development.

In theory there are four possible outcomes:

full Norwegian jurisdiction;

Soviet-Norwegian bilateralism;
internationalisation along the lines of the Spitzbergen Treaty;
no economic activity.

The first and the last solutions are in Norway's interest, as Norway will obviously have no economic need for producing oil here in the foreseeable future. The first, the second and the last solution could be in the interest of the Soviet Union. The first and the third solution could be in the interest of Norway's allies.

From a Norwegian point of view, economic neutralisation could be preferable. But it is unlikely that Norway would be able to take over large areas with promising prospects for petroleum and do nothing with them. It is a paradox, and a problem of Norwegian policy, that the restrictive attitude towards oil development is accompanied by an explicit desire to control large maritime areas. The Norwegian arguments are that the fisheries must be protected and that a restrictive oil policy also guarantees Western Europe future supplies of oil. These arguments are valid, but with the prospect of a more chronic scarcity of oil, it is unrealistic to assume that Norway will be able to withhold promising prospects of petroleum from the world economy. Consequently it is also in Norway's own interest to find an equitable solution.

The problem of the northern waters is briefly discussed in the 1974 oil plan. Here it is stated that it is the intention of the government to secure a firm Norwegian leadership of the oil activities in the northern-most waters.[18] It is also assumed that Norwegian national control in these areas will also be in the interest of the great powers. The most equitable solution could be to give Statoil full control of oil activities here.[19]

Notes

1. *World Energy Outlook* (OECD, Paris, 1977), p. 54.
2. Ibid., p. 28.
3. *Petroleumsvirksomhetens plass i det norske samfunn*, St.meld.nr. 25 (1973–4), (Ministry of Finance, Oslo, 1974), p. 87.
4. *Petroleumsvirksomhetens plass i det norske samfunn*, p. 87.
5. John Ausland, *Vår olje – vår makt?* (Cappelen, Oslo, 1978), p. 30.
6. Ausland, *Vår olje*, pp. 31 ff.
7. Nicholas Sarkis, *Le pétrole à l'heure arabe* (Editions Stock, Paris, 1975), pp. 208 f.
8. Øystein Noreng, 'Norges olje i Nordens energibalanse', *Internasjonal Politikk*, no. 2B (1975), pp. 353–71.

9. Lawrence G. Franko, *The European Multinationals*, (Harper and Row, New York, 1976), pp. 45 ff.

10. Jens Evensen, 'Den nye havrettsorden' in *Norges havretts − og ressurspolitikk* (Tiden Norsk Forlag, Oslo, 1976), pp. 14−42.

11. Finn Sollie, 'Det Utenrikspolitiske Bilde: Om oljeutvikling og utenrikspolitisk utvikling i Norge' in Thomas Chr. Wyller and Kari Bruun Wyller (eds.), *Norsk Oljepolitikk* (Gyldendal, Oslo, 1975), pp. 175−95.

12. *Petroleumsundersøkelser nord for 62° N*, St.meld.nr. 91 (1975−6), (Ministry of Industry, Oslo, 1976), pp. 30 ff.

13. *Oil and Security*, a SIPRI Monograph (Almqvist and Wicksell, Stockholm, 1974), p. 18.

14. *Petroleumsundersøkelser nord for 62° N*, p. 33.

15. John Ausland, *Vår olje*, pp. 98 f.

16. Finn Sollie, 'Det Utenrikspolitiske Bilde', pp. 175−95.

17. Kaare Sandegren 'Om Norges Sikkerhet og Havrettspolitikken' in *Norges Havretts − og Ressurspolitikk*, pp. 156−93.

18. *Petroleumsvirksomhetens plass i det norske samfunn*, p. 90.

19. Ausland, *Vår olje*, pp. 103 ff.

10 FUTURE PROSPECTS

The Record

In the late 1970s the record of UK and Norwegian oil policies appeared
in a somewhat different light from what was usually the case until a
few years ago. Norwegian oil policy has since the 1960s been described
as clear and consistent, successfully pursuing well defined aims. By
contrast, UK oil policy traditionally appeared less well managed.[1] By
1980, however, Norwegian oil policy appeared to have less clarity and
consistency, and to be managed with less skill and foresight. This is
partly connected with the problems of Norwegian industrial policy
and economic policy, and with the temptation to use oil policy as a
balancing factor to offset failures in industrial policy and economic
policy. But the management of the Norwegian oil sector itself also
appeared to be suffering from serious weaknesses. By contrast, UK oil
policy by 1980 appeared as relatively more successful, giving stronger
indications of skill and foresight than previously thought.

This is partly related to differences in levels of ambition and points
of departure of the oil policies in the two countries. Norwegian oil
policy was from the outset elaborated with much ambition, aiming at
full control of the activities. UK oil policy was initially elaborated with
fairly modest ambitions, aiming simply at the maximum economic
benefit. Consequently, even moderate problems in Norwegian oil
management can appear as dramatic failures, and modest successes in
UK oil management can correspondingly appear as more important
than they really are. But this more sober appreciation of records can
more easily be made from outside. In the UK and Norwegian contexts,
the changes in relative oil policy performances are often felt more
dramatically.

Given the favourable point of departure, the initial success of
Norwegian oil policy is not very remarkable. Also, given the highly
unfavourable point of departure, the initial mediocrity of UK oil policy
is not very remarkable either. It is more noteworthy that Norway
managed to manoeuvre herself into a fairly undesirable situation, largely
because of her own policies, and that the UK has managed to move
into a slightly more comfortable situation, at least for some time. This
has to do with the fact that the initially comfortable situation of the

Norwegian government in relation to oil permitted it to take a
multitude of different considerations into account, and thus to create
policy traps for itself, in which was subsequently caught. The initially
uncomfortable situation of the UK government left few choices open,
and there were few possibilities of creating policy traps, as the goal of
getting production going was predominant.

As pointed out in the first chapter, the aim of the governments was
both to capture a large part of the economic rent from oil and to have
control of the activities. These aims seem to have been pursued with
only moderate success. In both the UK and Norway the effect of the
oil taxation systems on the income distribution between governments
and companies is less to the benefit of governments than originally
intended. The recent taxation increases are motivated by this fact. In
addition, the taxation systems to a large extent make the cost escalation
a liability of the governments rather than of the companies. The aim of
control had two dimensions, the control of the entire oil activity and
the control of the companies' behaviour. At neither level is the success
very remarkable – production has been delayed and costs escalate,
and there are still evident problems of making the companies
conform to social and environmental regulations.

It is an open question whether, measured against the criteria of
capturing economic rent and securing industrial control, the UK's
performance has not been the most successful one. UK oil taxation
evidently has given a slightly higher total government take than the
Norwegian one. In addition, the cost escalation is in many ways a
more serious problem for Norway than for the UK, with a more
negative effect on public revenue.

The British have been more successful in getting the production of
oil and gas going, not only in absolute terms but also up to the levels
desired and projected. This also means that the UK government has
been more successful thant he Norwegian government in getting the
projected revenues and the projected impact on the trade balance.
Also, in recent years the UK government has obviously been more
successful than the Norwegian one in securing an industrial spin-off
from the oil activities. The Norwegian record is better when it comes
to social conditions in relation to the oil industry on land and at sea.
But in the Norwegian context requirements and ambitions were
higher as well. Furthermore, it is quite possible that the Norwegian
protection of the environment and safety precautions in relation to

the oil industry are the better ones, but Norway also had much higher
initial requirements and ambitions, and the most serious accident so
far has taken place in Norwegian waters, revealing insufficient
preparedness and control.

UK and Norwegian oil policies should be seen critically at two
levels, at the level of policy formulation and at the level of policy
implementation. Corresponding to national administrative traditions,
Norwegian oil and related policies have been formulated in a rather
ambitious and sophisticated way, emphasising theoretical argument
and extensive documentation, and UK oil and related policies have
generally been formulated with much less ambition and sophistication,
emphasising practical argument without extensive documentation. This,
of course, means that the standard by which the experiences are
judged differ, and that they tend to be considerably harsher in Norway
than in the UK.

By contrast, also corresponding to national administrative traditions,
the implementation of oil and related policies in Norway has not been
very active, and supervision not very detailed, and much has been left
to the trust that the oil companies, like other companies operating in
Norway, would willingly and loyally behave according to government
policies, whereas the implementation of oil and related policies in the
UK has been active, with detailed control and supervision. A typical
example is provided by the public administration concerning the use of
domestic goods and services. The Norwegians were the first to raise
this point, and in the Norwegian Oil and Energy Ministry there are
perhaps fair bureaucrats who look after this aspect of policy. By
contrast, the British, who more belatedly raised this point with the oil
companies, have the Offshore Supplies Office, where several hundred
people are working.

Certainly for Norway, and to a large extent for the UK, the task of
organising and controlling the oil industry and integrating it into the
national economy has been the greatest challenge to the governments in
time of peace in this century. Given the administrative traditions of
Norway, the Norwegian civil service has set high standards by which to
judge its performance, without having the means to accomplish a task
that is much more complex than the other tasks with which it has been
confronted. By contrast, also given local traditions, the UK civil
service has, wisely enough perhaps, not set very high standards by
which to judge its performance, but had the means to accomplish fairly
complex tasks. In this respect the critical variables are detail control
and information.

The difference in detail control also implies a difference in information feedback. Given a fairly detailed supervision, the UK government could be fairly well informed of the different aspects of the oil industry, for example as concerns the build-up of production. In Norway, the lack of detailed supervision meant that the government was ignorant to a certain extent about critical aspects of the oil industry, such as build-up of production, cost escalation, social aspects and the protection of the environment.

In the UK, the need to get production going as soon as possible together with detailed supervision has permitted a certain success along this dimension. In Norway, the initial relative freedom of action, together with a subsequent lack of detailed control, has permitted the government to be caught in its own policy traps. As already pointed out, it can be a wise policy to restrict oil development in order not to overheat the domestic economy, and it can be another wise policy to use anticipated oil revenues to offset an international slump. But the wisdom of combining these two policies is questionable, and such a combination does constitute a trap, because the spent anticipated revenues can be fairly high in relation to the real ones. In any case, such a policy does require fairly close control and supervision of the critical variables, and this has evidently been lacking. Similar problems exist in a number of other areas.

Against this background there is at present a fair amount of satisfaction with the oil policy in the UK, and this increasing consensus is illustrated by the fact that the key aspects of the oil policy have been retained by the present Conservative government. In Norway, on the contrary, there are tensions building up around the oil policy. On the one hand, there is an increasingly vocal build-up of dissatisfaction with the performance of foreign oil companies and with the inadequate government supervision. The dissatisfaction concerns the delays in production build-up, cost escalation, the public revenue and the inadequate participation by Norwegian industry as furnishers of goods and services, as well as social conditions and the protection of the environment. In public opinion and in the government there is a growing feeling that control over the oil industry is inadequate, and even a suspicion of being deceived by the international oil industry. This is a good argument for preferring national companies, especially Statoil, in the future, particularly because the record of the foreign companies in keeping cost and time targets is fairly unsatisfactory. On the other hand, the increasing combination of oil and industrial policy could imply a certain erosion of the principle of national control by opening up Norwegian waters for more foreign companies, and reducing rather than improving the potential for control. This conflict of interests could become more acute, affecting relations between government departments and within the Labour Party. Against this background, Norwegian oil policy seems to be approaching an important crossroads, particularly with the commencement of oil activities off northern Norway. The choices are either a movement in the direction of full nationalisation of the oil industry or a movement in the direction of an increasing reprivatisation of the Norwegian oil industry.

The record, particularly in Norway, but to a less extent in the UK as well, indicates that the North Sea model, at least as practised so far, does not function too well from the government's point of view, compared with policy goals. There is the problem of control, where the concessionary system as applied in UK and Norwegian waters, giving the formal right to organise operations and to dispose of the oil, has to be offset by government regulations for social and political reasons. This creates a permanent tension between the interests and the routines of the private companies and the aims and regulations of the state. The result so far is not satisfactory for the governments. It is doubtful whether it will remain satisfactory for the oil companies. In addition, there is the problem of the economic rent. In both countries, and particularly in Norway, the government take seems to be less, in some cases a good deal less, than what appears to have been agreed on in 1974–5. This means that the return on private capital invested in UK and especially Norwegian oilfields is currently a good deal higher than what was considered necessary and equitable in the mid-1970s.

As the current practices seem to give a fair degree of satisfaction neither on the issue of control nor on the issue of capturing the economic rent, alternative solutions should be considered. One solution could be to opt for licensing by auctions, in order to capture the maximum economic rent, and to give private companies a larger role. Taxation and other regulations could be kept more or less intact. An incentive for the companies to respect government regulations could be to exclude from new licensing companies which have a bad record in observing safety regulations, social regulations, etc. The advantage of such a solution would be to disengage the government from the oil operations and to put more distance between the civil service and the oil companies, so that in theory the civil service could perform its task of supervision and control more satisfactorily. Another advantage could be to attract more private capital, and to engage less public capital in the ventures.

Another solution could be to discontinue the concessionary system altogether, in order to achieve full government control and to reduce the role of private companies. The system of taxation and other regulations, historically designed for coping with the behaviour of private companies, would have to be changed. Oil taxation could be replaced by revenue sharing, or by channelling all oil revenues to the Treasuries. Safety and labour regulations could be implemented more directly by the responsible government agencies. The advantage of such a solution would be to improve controls, to make the civil service more directly responsible and make supervision more direct.

The disadvantage of the first solution is that it would drastically improve the position of private oil companies in relation to government, and that the government would be deprived of information and important policy instruments. This, in turn, might make both controls and capture of the economic rent more difficult in practice.

The disadvantage of the second solution is that the oil operations would rely more exclusively upon the state oil companies, which would then get a *de facto* position of monopoly, and that the oil operations would have to be financed more exclusively by public capital. This, in turn, might also make controls and capture of the economic rent more difficult. It could be argued that, in practice, a state monopoly for oil production, or for incremental oil production, would mean the expansion of the power and wealth of the state oil company rather than of the state. Thus, the nationalisation of oil might have results quite different from the policy aims envisaged. In this regard it can be pointed out that the position of monopoly of the state oil companies can be modified by attaching other oil companies to the operations, through service contracts, entrepreneurial contracts, etc. It can also be pointed out that the state has important policy instruments in relation to state oil companies, of which the possibility of centralising oil revenues to the Treasury and imposing a budgetary control on the state oil companies should be emphasised again. Finally, it should be pointed out that the trend towards oil nationalisation is world-wide, and that it seems to work fairly well in most places.

These considerations are perhaps more relevant for Norway than for the UK. In the UK context, with the prospect of declining oil production and rising imports of oil after 1990, it can appear as unnecessarily cumbersome to change the oil regime. In Norway, by contrast, with the prospect of increasing oil production in the opening up of the northern waters, it can appear as quite necessary to review the whole question of the oil regime. There are currently strong pressures in the direction of greater private involvement. This is a direct result of the recent policy to exchange oil for jobs, which allows foreign companies to gain access to licences as compensation for creating employment in Norway. An important consequence of this policy is not to make full use of the option of state participation, and consequently to let private interests have a greater share than would otherwise have been the case. In this respect it is symptomatic that the recent proposal to start oil operations in the northern waters explicitly states the need to let 'experienced foreign companies with sufficient capital' take part in the operations.[2] This can to some extent be seen as breaking the principle set in 1973–4 that oil operations in the northern waters should be carried out under firm Norwegian control.[3] However, it can be seen as consistent with the recommendation of the International Energy Agency to 'ensure that qualified foreign entities are allowed to participate in future hydrocarbons activity'.[4]

Given Norway's political history, in the 1970s as well as earlier in this century, it is unlikely that any foreign pressure systematically to influence Norwegian oil policy, and in particular to diverge substantially from the pattern set in 1973–4, relying upon an increasing degree of state participation, will go unopposed. It is likely that open foreign pressures in the direction of a higher level of output and increasing participation by foreign oil companies would trigger off serious opposition. Possibly this could lead to a polarisation of public opinion over oil policy along lines similar to those of the EEC conflict in the early 1970s. The outcome might not be favourable to the presence of foreign oil companies, or to an increase in the level of production. Such a conflict could also have important repercussions on foreign policy and on Norway's links with her Western allies. In the Norwegian political context the other alternative, of gradually moving towards a replacement of the concessionary system by a system of service contracts or entrepreneurial contracts, and thus gradually nationalising the oil activities, will probably be less divisive. Consequently, a gradual increase in the level of output might be politically more acceptable if it was taking place together with increasing state participation. These observations should be relevant for the oil consuming countries receiving Norway's oil and gas and wanting increased deliveries to offset declining oil production and insecure supplies in other parts of the world.

Notes

1. *The Economist*, 26 July 1975, North Sea oil survey, p. 6.

2. *Petroleumsundersøkelser nord for 62° N*, St.meld.nr. 57 (1978–9), (Ministry of Oil and Energy, Oslo, 1979), p. 81.

3. *Petroleumsvirksomhetens plass i det norske samfunn*, St.meld.nr. 25 (1973–4), (Ministry of Finance, Oslo, 1974), p. 8.

4. *Energy Policies and Programmes of IEA Countries, 1977 Review* (OECD, Paris, 1978), p. 14.

11 CONCLUSION: A QUESTION OF LEARNING

The Experience of Governments

During the period reviewed in this book, the 1960s and 1970s, the relationship between oil companies and governments in the UK and Norway has changed significantly, with the bargaining position of the governments gradually improving at the expense of the oil companies. This changing relationship is essentially due to two sets of factors:

> internally, by the experience of the UK and Norwegian governments since the early 1960s;
> externally, by the changing relationship between oil exporting and oil importing countries in the 1970s.

Even if the UK and Norway are among the world's youngest producers of oil, they are no longer infants in this respect. By 1980 the UK and Norwegian governments have been exposed to issues raised by the presence of the international oil industry for almost 20 years. This has been a period of trial and error, but it has also been a period of learning. The experience acquired in dealing with the oil companies and the knowledge that there is petroleum to be found and produced would in any case have improved the bargaining position of the governments significantly. Consequently, it is likely that certain changes in the initial regime, essentially the introduction of state participation, were bound to come, even if external circumstances had not changed dramatically. This was demonstrated by the modifications of the Norwegian oil regime prior to the 'oil revolution'.

However, it is difficult to understate the impact of the 'oil revolution' of 1973–4 on the bargaining position of the UK and Norwegian governments in relation to the international oil industry. To some extent, the impact may not yet have been fully realised, which could leave room for future changes of oil policy. To many observers, the fourfold price rise in a matter of months seems to overshadow the importance of the nationalisation of the oil industry in most OPEC countries, which created a major structural change in the world oil market, decoupling the mechanisms of supply from those of demand. The price rise made oil produced in UK and Norwegian waters much more profitable than before. The oil nationalisation made the access to equity oil in concessionary areas and access to secure supplies of oil in general much more valuable than previously.

The UK and Norway have not followed the example of most OPEC countries of nationalising existing oil concessions. One reason is that the relationship with the oil companies that existed prior to the 'oil revolution' was less unfavourable to the UK and Norwegian governments than was the case for most OPEC governments. Another reason is that the UK and Norway were more bound in their dealings with the oil companies than were the OPEC countries because of their close economic and political links with the industrialised oil consuming countries, primarily the United States.

On the other hand, UK and Norwegian practices have also diverged from those of the United States. State participation, levels of taxation and in Norway at least depletion policy distinguish UK and Norwegian oil regimes from the concessionary system as practised in the United States. One reason is that there has been fairly persistent political pressure in both countries in the direction of stiffer terms for the oil companies. Another reason is that such a stiffening of terms has been rational, given the objectives pursued by both countries' governments. A third reason is that governments have learned something from dealing with the oil companies over a long period of time.

Even if many of the oil companies' operations are still undisclosed to outsiders, the oil industry no longer appears as the enigma to the UK and Norwegian governments that was the case in the early 1960s. The micro-economic realities that distinguish the oil industry, and traditionally make it difficult to control from outside, are now largely known to the UK and Norwegian governments, partly because of their own state oil companies. This does not mean that the two governments are able to supervise all operations of the oil companies, but they have over time acquired a fairly comprehensive picture of how the oil industry works.

Consequently, it can no longer be assumed that the two governments do not know how to handle the oil companies. This also means that grave mistakes in oil policy should no longer be attributed to lack of experience and knowledge, but rather to bureaucratic mismanagement and explicit political preferences. In this respect, the responsibility of government is increasing. Correspondingly, it is increasingly erroneous to blame the oil companies alone when things go wrong. For a fairly long time both governments have been exposed to the development of oilfields on their continental shelves and to the task of securing national industrial participation. Both countries have for several years been exposed to oil revenues, to some extent indirectly through external borrowing anticipating future oil revenues. Thus, both the problems of securing for local industry a fair share in the development and the problems created by increasing liquidity through oil revenues should be known to governments. Consequently, when local industry does not get a fair share of the oil development and when the increased freedom of action in macro-economic policy that is based on oil revenues does not have entirely beneficial effects, governments and politicians are essentially to blame.

The learning process since the early 1960s has enabled the two governments to operate on a less unequal basis with the oil companies. Thus, governments have been increasingly able to ensure that salient objectives of oil policy are followed up in practice. In some cases this has required fairly strict measures. An important aspect of the learning process is that the two governments are realising that the oil companies, even after the 'oil revolution' when the bargaining position has changed considerably, are difficult to control from outside. The oil industry is still characterised by capital intensity, high profits potential, vertical integration and joint ventures, the very features that historically have been the basis of its independence and its shield against outside interference. It is also being realised that on matters such as depletion policy, field development strategies and use of local goods and services, there is a genuine conflict of interests between the governments and the private international oil companies, where the latter for reasons of their own will tend to have different objectives and priorities from the governments. Another aspect of the learning process is that the governments are realising the difficulty of changing rules and laws once they are set, and that government policy has a limited flexibility compared to that of the oil companies, and compared to the requirements of a constantly changing oil market. This is especially valid for taxation policy, which takes a long time to change, giving a distribution of income between the oil companies and the governments that fluctuates considerably with the prices in the market, and that the governments can only afterwards catch up with. Thus the oil companies are able to count on substantial additional profits for the period between a price rise in the oil market and the enactment of new tax legislation. This inertia represents a net loss of income for the governments. Within the framework of the concessionary system this is difficult to avoid.

The learning process is currently leading to a critical reassessment of experiences and policies by governments and politicians. The change of attitude by the new UK Conservative government, retaining the system of state participation and implementing tax changes proposed by its Labour predecessor, is remarkable. It indicates that the intention to secure a large part of the oil rent for the state is no less strong under the Conservatives than under Labour. It also indicates that the solutions prepared by Labour were relatively moderate. In Norway, there is a gradual realisation with both government and politicians that the national oil policy is much less perfect than previously thought. There is also a realisation that it is difficult for a small country with a limited industrial base to cope with large international oil companies, whatever the rules are. In the Labour Party this is leading to a critical reassessment of the entire concessionary system, favouring increasing state participation, if not full nationalisation of future discoveries. Among the non-socialist parties there is an ideological bias against increasing state participation and nationalisation, but the attitude of local industrialists seems to be increasingly critical of the international oil companies. This could strengthen the case for reduced foreign and private participation in future discoveries.

Experience of the Oil Companies

The petroleum activities in the North Sea have provided the oil companies with a unique experience in exploration, field development and production in conditions with deep water and heavy weather. The North Sea has been a test ground whose lessons are valuable for similar development elsewhere, first of all in Arctic and sub-Arctic areas. To some extent the oil taxation systems of the UK and Norway have permitted extensive experimentation with government subsidies, given the liberal provisions for cost deductions. The North Sea has also been a test ground for oil activities in a complex economic and political environment. The oil companies have had to learn to adapt to increasingly stiffer terms imposed by governments in matters of taxation, protection of the environment, depletion policy, use of local goods and services, labour, etc. Their presence in UK and Norwegian waters has also provided the oil companies with an experience of what issues could be raised politically, and of how the political systems would work out solutions that would not always be rational from the point of view of the oil industry, but that would strike a balance between different political demands. Against this background, the oil companies, for the sake of their own future in the UK and Norway, have had to learn to co-operate more closely both with governments and with local interests. It is fair to assume that this learning can be valuable for the activities of oil companies in other parts of the world.

In North America there has been increasing politicisation of energy issues, especially oil issues, in the 1970s. An increasing number of issues are raised politically, generally leading to reduced freedom of action for the oil companies, and possibly to reduced output of energy. There is a parallel between the politicisation of the oil issues in North America and the multitude of demands presented in relation to the oil industry by different participants in the political processes of the UK and Norway. To some extent, the politicisation of the oil issues in North America seems to be more accentuated and more diversified than is the case in the UK or Norway. This has to do with the different nature of political parties and the more fragmented political processes in North America. Furthermore, the UK and Norway have long records of co-operation of private and public interests, and of co-ordination of private and public decisions, that so far seem to be lacking in North America, especially in the United States. This means that the processes of politicisation of oil issues in the UK and Norway are more structured than is the case in North America, and that the mechanisms of conflict management in relation to oil are more efficient. Whereas it would be difficult for the oil companies to change the political processes in the United States, the experience gained by the consultative relationship with the UK and Norwegian governments might be of some relevance for the North American scene also.

So far, the oil companies have been fairly able to cope with changing political and administrative circumstances in UK and Norwegian waters. This might indicate that the changes of conditions implemented by the two governments are not radical, leaving the oil companies considerable freedom of action as well as substantial profits. It also indicates that the international oil companies are flexible and efficient organisations that are fairly able to adjust to changing external circumstances, and to adapt to new political conditions. In this perspective it is fair to assume that the international oil companies would also be able to adjust to a possible departure from the concessionary system, for example in Norway, and to use successfully a system of service contracts and entrepreneurial contracts that would give the government, or the state oil company, more sovereignty in micro-economic matters. Consequently, the international oil companies are likely to remain useful partners for governments even under this set of circumstances. The international outlook is that oil is likely to become increasingly expensive and increasingly scarce, which makes secure supplies of oil particularly attractive. This is the strength of oil producing governments; but it is also the strength of oil companies that have established relations with oil producing countries. There is little doubt that the international oil companies are learning this. The flexibility of the oil companies compared to that of governments indicates that they may not be the losers that they appear to be through the termination of the concessionary system in many parts of the world.

The Political Change

As underlined several times in this book, the relationship between oil companies and governments has changed throughout the world in the 1970s. It is within this context that the UK and Norwegian oil experiences should be analysed. The nationalisation of the oil industry in the OPEC countries and the current trend towards state trading in oil, on the basis of state-to-state deals, has politicised the international oil market. This has changed the relationship between oil exporting and oil importing countries, as well as between oil companies and governments within the different countries. Broadly, the oil exporting countries have drastically improved their bargaining position with the oil consumers and most governments have realised that oil is too important to be left to the oil industry alone. In many ways the 'oil revolution' was the event that signalled the end of the post-1945 world, and especially the breakdown of the institutional framework of the international economy conceived at the end of the Second World War and dominated by the United States, the Bretton Woods system. But this breakdown was not caused by the 'oil revolution' alone, as the system had been under serious strain for a number of years, and the international financial and monetary problems as well as the inflation problem were grave before the 'oil revolution'. Because of the gradual disintegration of the Bretton Woods system and because of the power of the OPEC countries, there is a new world economic order slowly emerging. It is characterised by a changed relationship between industrial countries and developing countries. It is also characterised by a reinforced role of the state in the different national economies, including the capitalist industrial countries. This is not confined to oil and other forms of energy. Increasingly the participants of the world economy appear as countries, as integrated economic-political systems where comparative advantages are efficient planning and systematic co-ordination of private and public decision processes. Within this international context the British and Norwegian oil experiences are not radical, and they fit parallel trends taking place in many other capitalist industrial countries. The British and especially the Norwegians were early to realise the need for state participation. Lately they have given private enterprise a larger share of the oil activities than practically all other oil exporting countries. The reason that there is more continuity in UK and Norwegian oil policies than in most OPEC countries is that the need for drastic change was less, because prior to the 'oil revolution' terms were less unfavourable to governments.

The British and Norwegian oil experiences throw an interesting light on the role of the state in developed capitalist democracies. Confronted with the issues related to a national oil industry, the UK and Norwegian states, at least under Labour governments, have acted contrary to both the prescriptions of orthodox neo-classical economic theory and to the anticipations of orthodox Marxist theory. Neo-classical economics would have prescribed that the task of organising a national oil industry be left to the market, i.e. to private enterprise, with a minimum of political interference. Orthodox Marxist theory would have anticipated that the state in capitalist society serves the interests of the bourgeoisie, i.e. the capitalist class, or its strongest faction, and facilitated its access to the most profitable parts of the economy. There is a complementarity between the prescriptions of neo-classical economics and the anticipations of orthodox Marxism, but neither seems to match well with the political realities of North-West Europe in the latter part of the twentieth century. In practice, it would definitely have been possible to leave the UK and Norwegian oil industry to the market forces and to private enterprise, with the state capturing part of the rent through taxation and auctioning off licences. Because there was no national industry able to assume the task in either country, this would have meant a heavy foreign dominance. Experience from other oil producing countries indicates that this would have created serious problems of control and potentially economic imbalances as well as political domination. In this perspective, the decision in favour of state participation, particularly in Norway, can be seen as the state operating on behalf of the capitalist class. Such an explanation would be compatible with orthodox Marxism. The point is, however, that when the state assumes control over one of the most profitable fields of economic activity, the balance of power between the public and the private sector is altered, as argued earlier in this book. This means that, even if the state should operate on behalf of the capitalist class, its doing so would also undermine to some extent the position of the latter.

Another explanation is that, in a capitalist society with a political democracy where representation is not confined to the capitalist class and where other social forces are well organised, the state is subject to a multitude of pressures and demands and has to strike a certain balance. This gives the management of the macro-economic framework a high priority, because it to a large extent determines the behaviour of the different participants in the political process, and because it may be easier to reach consensus on certain macro-economic targets than on micro-economic issues. However, this makes it difficult for the state to rely entirely upon market forces and private enterprise in issues of national importance. It also makes it difficult for the state to defend unilaterally the interests of the capitalist class or its strongest faction. In this way, in capitalist democracies there is a victory of politics over economics that had not been foreseen by neo-classical economics or orthodox Marxism. The maturity of the UK and Norwegian political contexts can be seen by the fact that the state to a considerable extent has had to withhold the most profitable field of economic activity from the capitalist class for the benefit of the state and wider social interests. The continuity of UK oil policy under a Conservative government is in this respect remarkable. It should also be underlined that, partly under the impact of anticipated oil revenues, the UK and Norwegian states have at times been fairly unable to secure the reproduction of private capital, and thus to perform the prime function ascribed to the state by orthodox Marxist theory.

Against this background it is reasonable to conclude that the UK's and Norway's becoming oil producing countries is equally a process of social and political change. The development of a national oil industry with the pattern of organisation chosen is a driving force in a process of social change. This process of change existed before the UK and Norway started out to become oil producers, and it is present in other European countries. However, the oil industry, because of its economic and political importance, tends to accelerate this process in the UK, and especially in Norway. Thus, the political economy of an oil exporting UK and Norway is quite different from that of the two countries in their pre-oil days. The balance between the private and public sector has been changed in favour of the latter. The counterpart to the extended role of the public sector is that it becomes more complex and more difficult to manage. In both the UK and Norway the process of change accelerated by the oil industry makes the task more urgent to organise the public sector of the economy in a way that is economically rational as well as socially responsible. This is also a process of learning.

INDEX

Milton Keynes UK
Ingram Content Group UK Ltd.
UKHW031146141024
449569UK00024B/1027